# Mathematical Questions from the Classroom Part I

# Mathematical Questions from the Classroom Part I

Richard J. Crouse

Clifford W. Sloyer

**Janson Publications, Inc.**
Providence, Rhode Island

Printed in the United States of America.

95 94 93 92 91 90 89 8 7 6 5 4 3

**Library of Congress Cataloging-in-Publication Data**

Crouse, Richard J.
Mathematical questions from the classroom.

Includes index.
1. Mathematics—Study and teaching (Secondary)
I. Sloyer, Clifford W. II. Title.
[QA11.C74 1987] 510′.7′12 86-27502
ISBN 0-939765-04-7 (pbk. : set)
ISBN 0-939765-02-0 (pbk. : v. 1)
ISBN 0-939765-03-9 (pbk. : v. 2)

# Contents

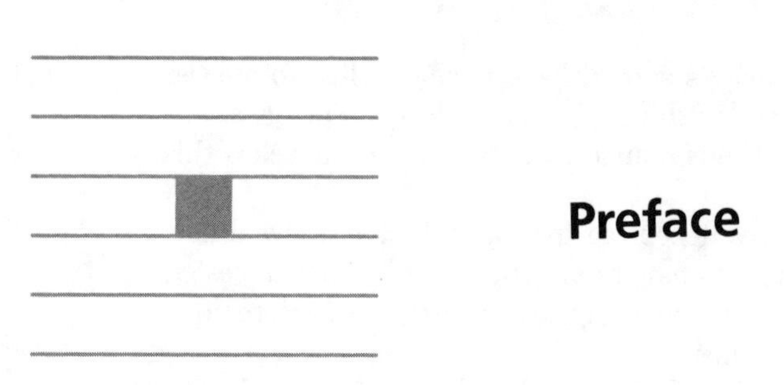

# Preface

After several years of teaching a methods course for college students preparing for secondary mathematics teaching, we found that these student teachers had great difficulty answering mathematical questions raised by their students. The authors began to collect questions that puzzled these university students as well as in-service teachers. This book is a compilation of those questions together with suggested solutions.

The majority of questions have come from actual classroom experience. We feel that these questions and answers are valuable resource material for teachers and can be used in a variety of pre-service, in-service, undergraduate, and graduate courses or seminars.

This printing of the book has two new features. First, the book has been divided into two parts. Part 1 consists of questions from Junior High School, Algebra I, and Geometry, while Part 2 consists of questions from Algebra II, Calculus, Elementary Functions, Probability, and Trigonometry. Second, the suggested answers for each section immediately follow the questions in order to make the book easier to use.

There are several ways in which this book can be used in class. One way is to tell the students that at the next meeting they will be asked questions on a particular topic. During the next meeting, questions from the given category are read and students are selected to answer them. One can either read the question and then select the student or select the student first and then read the question. Another way to use this material is to tell the class that they will be asked questions without giving them the topic. Then proceed as above. If a student has difficulty, invite help from others in the class. In either case, the students practice "thinking on their feet" and this emphasis allows the instructor to focus discussions in many directions, for example, how this concept might be taught in schools and how it is related to other topics. Some of these questions are difficult to answer and may be given as homework assignments to be discussed later in class.

Questions are organized by subject (e.g., Algebra I, Algebra II, Geometry, etc.) and are mainly of two types. Type I consists of those questions teachers usually face when teaching a specific topic. Type II consists of those questions that occur occasionally, but are the type many mathematics teachers would have difficulty answering.

One of the qualities we feel all teachers should have is the ability to answer content questions from their students while "thinking on their feet." The questions in this book have been used in our methods course to help students develop this competence.

For many of the questions included, a dozen teachers would offer twelve different responses. We make no claim that our answers are the "best." The suggested answers are simply a means of stimulating teachers and future teachers to look constantly for better pedagogical techniques.

We would like to express our thanks to the people who assisted us in the preparation of this text. In particular, we would like to thank the many student teachers and practicing teachers of mathematics who supplied us with many of the questions. We also particularly appreciated the constructive criticism offered by reviewers, Willard Baxter (University of Delaware), Margariete Montague (Northern Illinois University), Robert Washburn (Southern Connecticut College), John G. Harvey (University of Wisconsin, Madison), and John Wagner (Michigan State University). We would also like to thank Cathy Chudzik, Alison Chandler, Dorothy Voorhees, Sarajon Martin, and our wives, Joann Crouse and Barbara Sloyer, for doing a magnificent job in typing the original manuscript.

Richard J. Crouse
Clifford W. Sloyer

# 1. Junior High

## 1.1 Absolute Value

### Question

1 We are often faced with some very hard content questions to explain even at the seventh and eighth grade levels. How would you help a student to answer the following question? "You said $|a| = -a$, but how can the absolute value of a number be negative?"

### Answer

1 The main difficulty here is with the notation. If $a < 0$, then $|a|$ = the additive inverse or opposite of a. Since a is a negative number, then the additive inverse or opposite of a negative number is positive. For example, if $a = -3$, then the additive inverse of $-3$ is denoted by $-(-3)$ and equals 3.

## 1.2 Decimals

### Questions

1 A student asks you if it is possible to get all of the real numbers between any two real numbers by finding the average of these two numbers and continuing the averaging process indefinitely. What is your response?

2 When a student asks you if the set of irrational numbers is closed under addition, what is your reply?

3 A student asks you the following question. "When multiplying decimals, why do we count off decimal places from right to left in each factor and then add them in order to find out where to put the decimal point in the answer?" How do you respond?

4 A student asks, "We learned that if the denominator of a fraction in base ten has a factor of 2 or 5 then the decimal numeral is terminating. Will the same idea be true in other bases?" What is your response?

5 A seventh grade student asks you, "Why when multiplying or dividing decimals do we put the decimal point where we do." What is your response?

6 You make the statement that $1.\overline{9} = 2.0$. Some of your students protest, saying that the above statement is impossible because you have nines all the time after the decimal point. How do you handle this situation?

7 One student asks you the following question. "Why isn't 0.12½ equal to 0.62, since ½ = 0.5 and 0.12 + 0.50 = 0.62?" How would you answer this student?

## Answers

1 No, it is not possible. For example, if the numbers 3 and 4 are chosen, then since the averaging process will involve only rational numbers, one would never be able to get $\pi$ or any other irrational number between 3 and 4 by this process. One would get rational approximations for these irrational numbers, but never the numbers themselves. One should emphasize that the set of rational numbers and the set of irrational numbers are disjoint.

2 The set of irrational numbers is not closed under addition. For example, $\sqrt{2} + (-\sqrt{2}) = 0$ and 0 is a rational number. Another example is the following: $(3 + \sqrt{2})$ and $(-\sqrt{2})$ are irrational numbers, but $(3 + \sqrt{2}) + (-\sqrt{2}) = 3$ is a rational number. One should emphasize that the set of rational numbers and the set of irrational numbers are disjoint sets.

3 One way to respond would be to take a specific problem and show why the strategy works as follows. Suppose one multiplies the following using the traditional algorithm. The result is given below.

```
   3.24
  X .35
   1620
   972
 1.1340
```

If one writes out each decimal as a fraction one would get the following:

$$3.24 = 3 + \frac{2}{10} + \frac{4}{100}$$

$$.35 = \frac{3}{10} + \frac{5}{100}$$

Thus

$$3.24 \times .35 = \left(3 + \frac{2}{10} + \frac{4}{100}\right)\left(\frac{3}{10} + \frac{5}{100}\right)$$
$$= \left(\frac{300 + 20 + 4}{100}\right)\left(\frac{30 + 5}{100}\right) = \left(\frac{324}{100}\right)\left(\frac{35}{100}\right)$$
$$= \frac{(324)(35)}{10^4} = \frac{11340}{10^4} = 1.1340.$$

One should notice that the exponent of the denominator is 4, which is the same number one gets by counting off decimal places in each factor and adding. Thus, the reason why the decimal algorithm works is because of the equivalency of the decimal and fractional notation in our numeration system.

4 The statement which the student made is imprecise. For example, the fraction 2/15 has a denominator with a factor of 5 but the fraction cannot be represented as a terminating decimal. The student meant to say the following: a fraction a/b (in lowest terms) can be expressed as a terminating decimal if and only if the prime factorization of b is of the form $b = 2^x \cdot 5^y$ where x and y are whole numbers.

The same idea can be generalized to other bases as follows. The fraction a/b (in lowest terms) in base N can be expressed as a terminating decimal in base N if and only if the prime factorization of b in base N is of the form

$$b = c_1^{x_1} c_2^{x_2} \cdots c_k^{x_k}$$

where the $c_i$, $i = 1, \ldots k$ are prime numbers (in base ten) and each $c_i$ divides N.

5 For an answer, see the explanation for Section 1.2, Answer 3.

6 For an answer to this question, read the following article by L.T. Hall.* Also read:

W. Peterson, "Beware of Persuasive Arguments," *The Mathematics Teacher,* December 1972, p. 709.

# PERSUASIVE ARGUMENTS: .9999... = 1

**By LUCIEN T. HALL, JR.**

**Richmond Public Schools**
**Richmond, Virginia**

WHEN teaching high school students, prospective teachers, and in-service classes, I have used the usual method for changing the repeating-decimal format of a rational number to an $a/b$ format where $a$ and $b$ are integers and $b \neq 0$. The usual

*From *The Mathematics Teacher,* December 1971, pp. 749-50. Reprinted by permission.

method referred to is, using .7272 . . . as an example,

| | |
|---|---|
| to let | $N = .7272\ldots,$ |
| multiply both sides by 100 | $100N = 72.7272\ldots,$ |
| subtract equals from equals | $99N = 72.0000\ldots,$ |
| and divide both sides by 99 | $N = 72/99.$ |

Thus, .7272 ... = 72/99.

Students readily accept this but balk when an example has repeating 9s—such as .9999 . . . or .4999. . . . Several arguments can be used to persuade students that, for example, .9999 . . . is exactly 1.

*First Argument.* Use the usual method:

$$N = .9999\ldots,$$
$$10N = 9.9999\ldots,$$
$$9N = 9.0000\ldots,$$
$$N = 9/9.$$

So, .9999 . . . is 1. Ask the students emphatically, "Have I made a mathematical mistake?" The usual answer from students is "No, but . . . ."

*Second Argument.* Tell students that this argument is no proof but good intuitive reasoning. Ask them to express .1111 . . . in $a/b$ form, express .2222 . . . in $a/b$ form, express .3333 . . . , and so on, and finally .8888 . . . in $a/b$ form. The students answer 1/9, 2/9, 3/9, and so on, and finally 8/9. Then say, "Intuitively speaking, then, what is .9999 . . . ?" Students acknowledge that the answer is 9/9.

*Third Argument.* Tell students that if .9999 . . . is not 1, then it must be either larger or smaller than 1. Which is it? Students always answer, "Smaller." So ask, "How much smaller?" This question is met with silence. You might tell students that if they think the number is smaller than 1 they must be thinking about a *large but finite* number of 9s rather than an infinite number of 9s in the decimal expression.

*Fourth Argument.* Ask students if they agree that 1/3 = .3333. . . . When they readily agree, then ask if you might multiply both sides by 3. Naturally you get 3(1/3) = 3(.3333 . . .), or 1 = .9999. . . . This seems to be the most persuasive argument.

*Fifth Argument.* Ask students if they agree that 4/9 = .4444 . . . and 5/9 = .5555. . . . When they agree, then ask if you might add equals to equals. Adding, we get 9/9 = .9999 . . . , or 1 = .9999. . . .

*Sixth Argument.* Sometimes a very good student will suggest that if .9999 . . . does equal 1 or 9/9, then you should be able to change 9/9 to .9999 . . . by dividing the denominator into the numerator. You must remind students that a division problem is correct when the quotient times the divisor plus the remainder equals the dividend. You must also admit that we usually require the remainder to be smaller than the divisor but that this is a matter of definition and not necessarily mandatory.

Start dividing.

```
   .9
9/9.0000
  8.1
   .9
```

Note that the quotient times the divisor plus the remainder equals the dividend. Go several steps further.

```
   .9999 . . .
9/9.0000 . . .
  8.1
   .90
   .81
   .090
   .081
   .0090
   .0081
   .0009
```

At any point, the quotient times the divisor plus the remainder is equal to the dividend; 9 divided by 9 is producing the quotient .9999. . . .

7 The main problem here is one of determining what the symbol 0.12½ means. If the student was correct, then this would mean that a numeral such as .125 would mean .12 + 5, which it certainly does not. The symbol 0.12½ means

$$\frac{1}{10} + \frac{2}{100} + \frac{1}{2}\left(\frac{1}{100}\right).$$

Thus $0.12\frac{1}{2} = \frac{1}{10} + \frac{2}{100} + \frac{5}{1000} = .125.$

It is quite unfortunate and, at times, quite confusing when similar looking symbols mean different things. For example,

$$2\frac{1}{3} = 2 + \frac{1}{3}, \quad \text{but} \quad (a)\frac{1}{2} = a \cdot \frac{1}{2} \quad \text{and} \quad 2a = a + a, \quad \text{not } 2 + a.$$

You might want to point these out to the student and tell him that one of the skills that he must develop is the ability to understand what meaning to attach to different symbols in mathematics.

# 1.3 Equality

## Question

1 In your eighth grade class, one of your students asks the following question. "Is there a difference between *equal* and *equivalent* in mathematics?" How would you answer him?

## Answer

1 Yes, there is a difference between *equal* and *equivalent* in mathematics. Usually equality is used to mean that whatever is on one side of the equals sign names the same object as that on the other side of the equals sign. Thus, 1 + 2 = 3 means that "1 + 2" and "3" are names for the same number. Thus we say that the expressions "1 + 2" and "3" are equivalent expressions, but the numbers that they stand for are equal. The expressions themselves are not equal since they are not identical expressions. Likewise, the numerals 5/2, 2½, 2.5 and 250% are equivalent numerals since they name the same number. Thus the numerals are equivalent but not equal; i.e., the symbol 5/2 is not the same symbol as 2½, etc.,but the numbers represented by these numerals are equal.

Equivalency and equality are related but not the same concept. Equivalency is in many instances a generalization of equality and usually means that objects share a common property or attribute. Thus if $A = \{1, 2, 3\}$ and $B = \{3, 4, 5\}$, then A is equivalent to B since there is a one-to-one correspondence between the elements of A and B. If $C = \{2, 3, 1\}$, then we say that A is equal to C since they contain exactly the same elements. Thus, in this illustration with sets, we can say that if two sets A and B are equal then they are equivalent; but the converse statement need not be true.

As another illustration consider the following equation $3y^2 + 5y + 4y + y^2 = 4y^2 + 9y$. If we replace y by any member of its replacement set we have a true statement. Thus the members of this equation are called equivalent expressions. For any value of y in the replacement set, the equivalent expressions represent the same number and thus the equation is called an identity. The expressions "$3y^2 + 5y + 4y + y^2$" and "$4y^2 + 9y$" are not equal expressions, i.e., the same expressions.

As a last example, consider the statements $p \rightarrow q$ and $\sim p \vee q$ where p and q are statements. These expressions are clearly not the same expressions. However, if a truth table is made for each expression, the truth tables are identical, i.e., equal, and thus the expressions are called logically equivalent statements.

Space limits us from considering other examples in detail, but it could be a very interesting exercise for the reader to go through secondary mathematics material to find many other different places where the idea of equivalence is used. The following is an incomplete listing to start you off.

Equivalent expressions
Equivalent equations
Equivalent inequalities
Equivalent rational expressions
Equivalent systems of equations
Logically equivalent statements
Equivalent triangles
Equivalent polygons
Equivalent vectors
Equivalent matrices
Equivalent polynomials
Equivalent fractions
Equivalence classes

The name *equivalent* comes from the notion that when objects are equivalent, this relation usually turns out to be an equivalence relation. In many instances the distinction concerns the difference between an object (equality) and the names of the objects (equivalence). For an interesting article concerning equality and equivalence consult:

Walter Sanders, "Equivalence and Equality," *The Arithmetic Teacher,* April 1969, pp. 317-22.

## 1.4 Exponents

### Questions

1 In response to your question of how to define $y^k$, a student says that y is multiplied by itself k times. How do you reply?

2 A student asks, "Why is $x^0 = 1$?" How do you respond?

3 One of the students asks you what $0^{-7}$ is. How do you respond and why?

### Answers

1 This student is incorrect. Suppose that $k = 3$, then by the student's definition $y^3 = [\{(y \cdot y) \cdot y\} \cdot y] = y^4$. It would be much better to say that y is used as a factor k times.

2 $x^0 = 1$ mainly for our desire to maintain for the zero exponent the laws for positive integral exponents, i.e., $x^a \cdot x^b = x^{a+b}$ where a, b are positive integers. When we enlarge the domain to include zero, then we would wish the same law of exponents to hold. Thus,

$$x^0 \cdot x^a = x^{a+0} = x^a.$$

Therefore, $x^0 = x^a/x^a = 1$, assuming that $x \neq 0$. If $x = 0$, then we encounter the undefined expression $0^0$. An intuitive way of showing that $x^0 = 1$ is plausible is as follows:

$$\begin{array}{ll} 2^6 = 64 & 3^6 = 729 \\ 2^5 = 32 & 3^5 = 243 \\ 2^4 = 16 & 3^4 = 81 \\ 2^3 = 8 & 3^3 = 27 \\ 2^2 = 4 & 3^2 = 9 \\ 2^1 = 2 & 3^1 = 3 \\ 2^0 = ? & 3^0 = ? \end{array}$$

3 $0^{-7} = 1/0^7$ is undefined. In general, $0^{-k}$, where k is a positive integer, is undefined.

## 1.5 Fractions

### Questions

1 A student asks, "Since we have defined multiplication as repeated addition, how is it possible that when you multiply two fractions you get a product smaller than either of the factors?" How do you reply?

2 A sixth grader asks why $3\frac{1}{3} \times \frac{3}{5} = 2$? What the student does not understand is that the answer, 2, is smaller than one of the factors, $3\frac{1}{3}$. How do you reply?

3 You put the problem $\frac{1}{5} + \frac{2}{3}$ on the board and ask one of your students to come to the board and demonstrate how to compute the answer. He writes the following: $\frac{1}{5} + \frac{2}{3} = \frac{3}{8}$. You tell him that it is wrong, but he gives you the following explanation. "If a baseball players is up at bat 5 times on the first day and gets one hit, and on the second day he gets 2 hits out of 3 times at bat, then altogether he has 3 hits out of 8 times at bat." How do reply?

4 One student asks the following question. "Can you divide fractional numbers by dividing the numerator of the top fraction by the numerator of the bottom fraction and likewise the denominator of the top fraction by the denominator of the bottom fraction?" For example,

$$\frac{14}{9} \div \frac{2}{3} = \frac{14 \div 2}{9 \div 3} = \frac{7}{3}$$

How do you reply?

5 One student makes the following assertion. "1/2 = 1/3 because my picture here shows it."

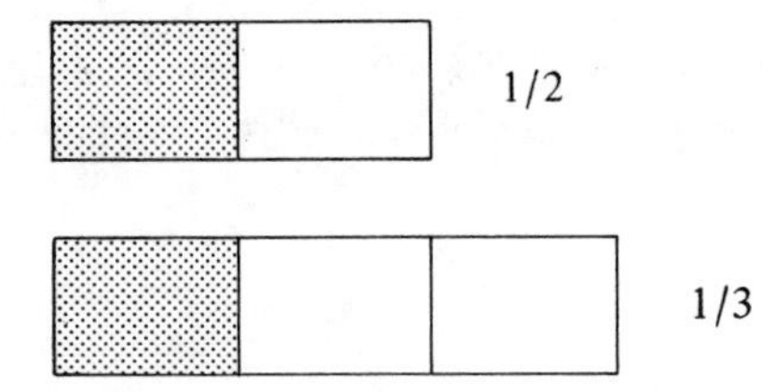


How would you explain to this student that he is incorrect?

6 A student says that there is something wrong with the definition of less than and shows you the following problem.

$$\frac{-3}{-4} < \frac{-2}{5} \quad \text{because} \quad (-3) \cdot (5) < (-2) \cdot (-4)$$

but $\dfrac{-3}{-4} = \dfrac{3}{4}$.

How would you answer this student?

7 An eighth grade student hands in the following solution when asked to reduce 64/16 to lowest terms.

$$\frac{\not{6}4}{1\not{6}} = \frac{1 \cdot 4}{1 \cdot 1} = \frac{4}{1} = 4. \quad \text{Is the student correct?}$$

8 There is ¾ of a pie left. Jane and Sue want to divide what is left in ½ so they each get an equal part. How much do each of them receive? A seventh grader's answer:

$$\frac{3}{4} \div \frac{1}{2} = \frac{3}{4} \times \frac{2}{1} = \frac{6}{4} = 1\frac{2}{4} = 1\frac{1}{2}.$$

What is your response?

9 You are teaching division of fractions. A seventh grader asks, "Why does multiplying by the reciprocal of the divisor give the correct answer?" What is your response?

10 A student in your class asks, "Why don't we need to get common denominators when multiplying fractions?" How would you respond?

11 In your seventh grade class, a student is doing the following problem.

$$\frac{3}{4} + \frac{5}{2}$$

The student writes $\frac{3}{4} + \frac{5}{2} = \frac{8}{4}$ and reasons as follows: "5 + 3 = 8 but since 2 will go into 4, then don't add the bottom numbers." How would you help this student?

12 A teacher asks a student the following question. "Which would you rather have, 20/35 of a cake or 4/7 of a cake?" The student says, "20/35 because there are then more pieces." How would you respond if you were the teacher?

13 You have given your class the following problem. "If the selling price of a car is \$3300 and the profit is 1/10 of the cost, then how much is the profit?" One student says that the profit is \$330 because \$330 is 1/10 of \$3300. Is the student correct?

14 An eighth grade student asks, "You said that a rational number is one of the form p/q, p, q integers, q ≠ 0. What about:

$$\frac{\sqrt{3} \cdot 3}{\sqrt{3} \cdot 4}\ ?$$

Is this a rational number or not?" How do you reply?

15 You have given your class the following problem to solve: "A 20-foot board is to be cut into shorter pieces. If each piece is 4¾ feet long, how many pieces will there be?" One student hands in the following work:

$$\text{“}20 \div 4\frac{3}{4} = 20 \div \frac{19}{4} = 20 \times \frac{4}{19} = \frac{80}{19} = 4\frac{4}{19}.$$

Thus there are 4 pieces and 4/19 of a foot left over." Is the student correct?

## Answers

1 When we write 2 · 3, we usually mean 3 + 3. Thus ¼ · ½ should be the number that is ¼ of 1 · ½ or 1 · ½ divided by 4. Thus the result is smaller than the original factor ½. Likewise, by the commutative law of multiplication ¼ · ½ = ½ · ¼ = ½ of 1 · ¼ or 1 · ¼ divided by 2. Thus, the result is smaller than the original factor ¼. In general, (a/b) · (c/d), (b, d ≠ 0; a/b, c/d proper fractions), can be viewed as

$$\left[\frac{1}{b}\left(a \cdot \frac{c}{d}\right)\right].$$

Thus c/d is first multiplied by a, and the result is divided by b. Since $a < b$, then the resulting number is less than c/d. Using the commutative property for multiplication, one can use a similar argument for a/b.

2 If we write $3\frac{1}{3}$ as the improper fraction $\frac{10}{3}$, then the answer to this question is given in Answer 1.

3 For an answer to this question, read the following article by Van Engen and Cleveland.*

# MATHEMATICAL MODELS FOR PHYSICAL SITUATIONS

By H. VAN ENGEN

University of Wisconsin
Madison, Wisconsin

and RAY W. CLEVELAND

Jersey City State College
Jersey City, New Jersey

RECENT attempts to improve the teaching of problem solving in mathematics have made it necessary to look carefully at physical situations and the way in which they are represented by conventional mathematics. Relatively recent interest in the newer studies in developmental psychology has focused attention also on the need for a closer examination of experiences being provided elementary and secondary pupils in connection with certain mathematical ideas as they relate to problem solving. There are instances in the physical world where it is possible, and at times desirable, to represent a physical situation by using mathematical ideas and their properties which are quite different from those commonly used.

Consider, for example, the two categories of physical situations that are popularly known as "ratio" or "fraction" situations. In these cases, two numbers are present, and no mathematical operation is implied.

*Category 1*

*a*) Three apples for five cents.
*b*) Three miles in five minutes.
*c*) Three inches to five inches.
*d*) Three marbles out of five are red.

*Category 2*

*a*) A pie is cut into five equal parts, and three parts are under consideration.

*b*) A ribbon is marked off into five equal parts, and three parts are cut off.

To teach children how the pairs of numbers are used in each of the two categories, one would ordinarily present entirely different physical arrangements of objects. In fact, a knowledge of Piaget's developmental psychology would cause one to question whether children consider the instances mentioned in the two categories as similar in any respect. Certainly, it is unwise to insist on calling all such instances "fraction" situations, although this is frequently done.

It is easier to see that the two categories are different if we attempt to define an op-

*From *The Mathematics Teacher*, October 1967, pp. 578-81. Reprinted by permission.

eration on the number pairs in each category. In Category 1, a purchase of three apples for five cents followed by a purchase of two apples for seven cents results in a total purchase of five apples for twelve cents. Using the symbolism for ordered pairs, we then have

$$(3, 5) + (2, 7) = (5, 12).$$

If we were to use the standard ratio symbolism, then we would write the following:

$$\frac{3}{5} + \frac{2}{7} = \frac{5}{12}.$$

In Category 2, three fifths of a cake and two sevenths of a cake, when considered as a total portion of one cake, result in $\frac{31}{35}$ of a cake, not $\frac{5}{12}$ of a cake.

$$\frac{3}{5} + \frac{2}{7} = \frac{21}{35} + \frac{10}{35} = \frac{31}{35}.$$

It is obvious that the rules for combining the pairs of numbers in Category 1 differ from the rules used in Category 2. The question now arises as to what mathematical model to use when the pairs of numbers are abstracted from each category. In Category 2, it is clear that the rules for computing with rational numbers enable us to find a satisfactory interpretation of the computational results. In other words, there is no deviation from commonly accepted practice. In Category 1, however, the pairs of numbers cannot be combined as if they were rational numbers if we are to interpret the result in a way that really represents the situation. This suggests that the rational numbers do not form a model for ratio situations. What, then, is the model?

In answering the question, we shall restrict our examples to pairs of integers. Furthermore, we shall use the symbols $\frac{2}{3}$ and (2, 3) interchangeably to represent the pairs of numbers as they occur in the physical situations of Category 1. First we shall give interpretations for pairs with negative components as well as pairs with positive components. These interpretations can be made if one is allowed the same liberties first-year algebra teachers demand when they interpret such statements as $(^-2)(^-3) = {}^+6$ for their students.

(2, 3) or $\frac{2}{3}$ means
- *a*) Get 2 apples, pay 3 cents (buying)
- *b*) Earn 2 dollars, put 3 dollars in the bank
- *c*) Move 2 miles north in the next 3 hours
- *d*) Take 2 reds, take 3 greens
- *e*) Go 2 miles east, go 3 miles north

(2, −3) or $\frac{2}{^-3}$ means
- *a*) Get 2 apples, get 3 cents
- *b*) Earn 2 dollars, withdraw 3 dollars
- *c*) Move 2 miles north in the past 3 hours
- *d*) Take 2 reds, give 3 greens
- *e*) Go 2 miles east, go 3 miles south

(−2, 3) or $\frac{^-2}{3}$ means
- *a*) Give 2 apples, pay 3 cents
- *b*) Give 2 dollars, put 3 dollars in the bank
- *c*) Move 2 miles south in the next 3 hours
- *d*) Give 2 reds, take 3 greens
- *e*) Go 2 miles west, go 3 miles north

(−2, −3) or $\frac{^-2}{^-3}$ means
- *a*) Give 2 apples, get 3 cents (selling)
- *b*) Give 2 dollars, withdraw 3 dollars
- *c*) Go 2 miles south in the past 3 hours
- *d*) Give 2 reds, give 3 greens
- *e*) Go 2 miles west, go 3 miles south

We now proceed to define and interpret an operation "+" on ordered pairs $(a, b)$ and $(c, d)$, where $a$, $b$, $c$, and $d$ are rational numbers. Remember that, in choosing our examples, we will use only integers.

*Definition:* $(a, b) + (c, d) = (a + c, b + d)$.

*Interpretation:* Two apples for three cents and five apples for eight cents is the same as seven apples for eleven cents.

$$(2, 3) + (5, 8) = (7, 11).$$

Obviously, the operation of addition of number pairs as defined above is closed, because $a + c$ and $b + d$ are rational numbers. Addition of number pairs has properties other than the closure property, and the properties are listed below.

$P_1$. *Closure:* $(a + c, b + d)$ is an ordered pair of rational numbers.

$P_2$. *Associative:* $[(a, b) + (c, d)] + (e, f) = (a, b) + [(c, d) + (e, f)]$.

$P_3$. *Identity:* $(a, b) + (0, 0) = (a, b)$.

$P_4$. *Inverse:* $(a, b) + (-a, -b) = (0, 0)$.

*Interpretation:* Buying two apples for three cents followed by selling two apples for three cents, or $(-2, -3)$, results in zero apples and zero cents or $(0, 0)$.

$$(2, 3) + (-2, -3) = (0, 0).$$

$P_5$. *Commutative:* $(a, b) + (c, d) = (c, d) + (a, b)$.

Next, let us define and interpret an operation of multiplying an ordered pair $(a, b)$ by a rational number $\alpha$.

*Definition:* $\alpha(a, b) = (\alpha a, \alpha b)$.

*Interpretation:* Two purchases of two apples for three cents is the same as a purchase of four apples for six cents.

$$2(2, 3) = (4, 6).$$

The properties of this multiplication are as follows:

$P_6$. *Closure:* $(\alpha a, \alpha b)$ is an ordered pair of rational numbers.

$P_7$. *Distributive:* $\alpha[(a, b)+(c, d)] = \alpha(a, b) + \alpha(c, d)$.

*Interpretation:* Two purchases of three apples for five cents and six apples for seven cents is the same as two purchases of three apples for five cents added to two purchases of six apples for seven cents.

$$2[(3, 5) + (6, 7)] = 2(3, 5) + 2(6, 7).$$

$$(\alpha + \beta)(a, b) = \alpha(a, b) + \beta(a, b).$$

*Interpretation:* $(4+6)$ moves, each move consisting of two units east and three units north, equals four moves of two units east and three units north, followed by six moves of two units east and three units north, as shown below:

$$(4 + 6)(2, 3) = 4(2, 4) + 6(2, 3).$$

$P_8$. *Associative:* $\alpha \cdot \beta(a, b) = \alpha \cdot (\beta a, \beta b)$.

*Interpretation:* Four repetitions of five purchases of two apples for three cents equals four purchase of ten apples for fifteen·cents, as shown below:

$$4 \cdot 5(2, 3) = 4(10, 15).$$

$P_9$. *Identity:* $1(a, b) = (a, b)$.

The properties $P_1$ through $P_9$ are those of a vector space, where the elements $(a, b)$ are pairs of rational numbers, and the elements $\alpha$ and $\beta$ are scalars, in this case rational numbers. The ordered pairs, which are the elements of the vector space, are called vectors. However, the physical interpretations given in this article are commonly regarded as ratios. What has been shown is that ratios satisfy a vector-space interpretation. In short, ratios are vectors.

There are a few difficulties in such an interpretation. For example, when people are using ratios in elementary mathematics, science, or business, they often write $\frac{a}{b} = \frac{c}{d}$ for $(a, b) = (c, d)$. It is well known that two vectors $(a, b)$ and $(c, d)$ are equal if and only if $a = c$ and $b = d$. Thus, if we give a vector interpretation to ratio situations, it is incorrect to write $\frac{a}{b} = \frac{c}{d}$ unless it is actually true that $a = c$ and $b = d$. A similar difficulty arises in connection with cross products of ratios; that is, if $\frac{a}{b} = \frac{c}{d}$, then $ad = bc$. How do these ideas fit, if at all, into a vector-space interpretation?

Suppose that we look at the graph of the following set of ratios: *a*) two apples for ten cents, *b*) four apples for twenty cents, *c*) six apples for thirty cents, and *d*) eight apples for forty cents.

The graph suggests that in this instance we are dealing with a subspace of a two-dimensional vector space. A linear subspace of a two-dimensional vector space is the set of vectors, $V$, defined as follows:

$$\{\alpha(a, b) \mid \alpha, a, \text{ and } b \in R\}.$$

This definition tells us that each vector in the same linear subspace as vector $(a, b)$ is a scalar multiple of vector $(a, b)$. Hence, the definition provides the clue for finding two vectors (ratios) that represent the same price. If the ratios represent the same price, they are in the same subspace. In solving problems like the following, therefore, we consider the ratios to be in the same subspace.

If two apples cost ten cents, how much will fifty apples cost?

We know that $(2, 10)$ determines the subspace in which we are interested; we want to find another ratio (vector) in the same subspace. Hence,

$$\alpha(2, 10) = (50, x).$$

$$(2\alpha, 10\alpha) = (50, x).$$

By our definition of equality of two ratios (vectors), we have

$$2\alpha = 50, \text{ or } \alpha = 25,$$

and

$$10\alpha = x, \text{ or } x = 250.$$

Hence, fifty apples cost 250 cents, or \$2.50.

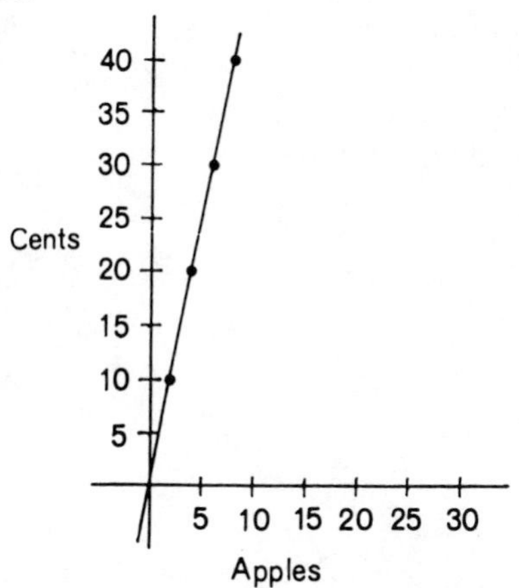


GRAPH OF $\{(2,10), (4,20), (6,30), (8,40), \ldots\}$

It is, however, more convenient to solve such problems by using a theorem that pertains to ratios in the same subspace; namely, the theorem that $(a, b)$ and $(c, d)$ are in the same subspace if and only if $ad = bc$.

The proof of the theorem has two parts.

*Part 1*

$\alpha(a, b) = (c, d).$

$\alpha a = c$ and $\alpha b = d.$

$\alpha ad = \alpha bc.$

$\therefore\ ad = bc$ if $\alpha \neq 0.$

*Part 2*

$ad = bc.$

$\frac{d}{b} = \frac{c}{a} = \alpha.$

$\alpha b = d$ and $\alpha a = c.$

$\therefore\ \alpha(a, b) = (c, d).$

This is the well-known cross product of two "equal" ratios. The fact that this property holds for two ratios in the same subspace and also for two rational numbers may be the source of the confusion that often exists concerning ratios and rational numbers.

Mathematics educators agree that the elementary and secondary schools must provide experiences that help students to understand the properties of the rational numbers. The rational numbers under addition and multiplication are an example of an important mathematical structure. We feel that the vector-space model is another structure that should be included in the school program. True, it is debatable whether elementary school children should be asked to write the following:

$$\frac{2}{3} + \frac{5}{6} = \frac{7}{9}$$

$$(2, 3) + (5, 6) = (7, 9),$$

but the question is still an open one. Certainly, program planners should be well aware of the problem confronting them with respect to ratios and rational numbers.

We now turn our attention to a second issue. Since the interpretation of certain pairs, such as $(-3, 4)$, is rather awkward in the situations usually presented to elementary and junior high school students, it might be desirable to limit ratios as vectors to those pairs $(a, b)$ for which $a > 0$ and $b > 0$; in other words, the vectors should be limited to the first quadrant of the plane. True, this limitation prevents one from exhibiting the entire vector-space structure, but this same limitation exists with respect to the rational numbers as taught in the elementary school. For example, the additive-inverse element is lacking in both instances.

In conclusion, we offer a few words about terminology. It makes little difference whether the pairs we have been considering are called ratios, rate pairs, or vectors. What an idea is called, after all, is only a name. Individual teachers may have various reasons for not liking a name, but these reasons are often personal, rather than mathematical or pedagogical. For instructional purposes, it matters very little what name is given to the pairs of numbers under consideration. Although a real case could be made for calling the pairs vectors, even in the elementary school, common practice in science and in the business world has to be considered.

To summarize, we have tried to show that ratios may be interpreted as vectors; that there are important curricular implications for a vector-space interpretation; and that there are instructional implications for this point of view. As a teacher, you should give the matter some thought.

4 Yes, you can.

$$\frac{a}{b} \div \frac{c}{d} = \frac{ad}{bc} \qquad b, c, d \neq 0$$

$$\frac{a}{b} \div \frac{c}{d} = \frac{ad}{bc} \cdot \frac{1/dc}{1/dc}$$

$$\frac{a}{b} \div \frac{c}{d} = \frac{ad \cdot (1/dc)}{bc \cdot (1/dc)}$$

$$\frac{a}{b} \div \frac{c}{d} = \frac{(ad/dc)}{(bc/dc)} = \frac{(a/c)}{(b/d)} = \frac{a \div c}{b \div d}.$$

Thus, $\frac{a}{b} \div \frac{c}{d} = \frac{a \div c}{b \div d}$.

5 The basic problem here is that the student is using different basic units for comparison purposes. The student must be made to realize that when comparing objects, the same units must be used in order to get a valid comparison. For example, if a student said that ½ = ¼, then one might ask this student if ½ of a dollar is equal to ¼ of a dollar.

6 The usual definition of less than for rational numbers is stated as follows: If a/b and c/d are expressions for rational numbers with b and d positive integers, then $a/b < c/d$ if and only if $a \cdot d < b \cdot c$. The definition requires that b and d both be positive, which is the condition violated in the student's example. The reason for requiring the condition is given as follows:

$$\text{If } \frac{a}{b} < \frac{c}{d}, \text{ then } \frac{ad}{bd} < \frac{bc}{bd}$$

as long as b and d have the same sign (i.e., either both are negative or both positive); otherwise the sign of the inequality changes. In order to ensure that this is always the case, then the condition follows since any natural number can be represented as a fraction with a positive denominator.

7 For an answer to this question, read the interesting article:

Fred B. Thomas, "A Dialogue on $2\not{6}/\not{6}5 = 2/5$," *The Mathematics Teacher*, January 1967, pp. 20-23,

8 The basic problem here is one of language. It probably would have been better to state the problem as follows: "There is ¾ of a pie left. Jane and Sue want to divide what is left in two parts so they each get an equal part." Even if the problem is left

as stated, the teacher might point out that the student's answer is wrong. How could each person get 1½ pies if they only started out with ¾ of a pie. Perhaps a good question to ask the student is the following: "How much should each of two people receive if they divide one dollar evenly between them?" The student should recognize that each person gets 50 cents and the student arrived at this answer by dividing by 2, not ½.

9 There are various ways of answering the student. Some of these ways are given below. A specific example is used in each case.

(a) $\frac{2}{3} \div \frac{4}{5} = r \leftrightarrow \frac{2}{3} = \frac{4}{5} \cdot r$ by the definition of division

$$\frac{2}{3} = \frac{4}{5}r \leftrightarrow \frac{2}{3} \cdot \frac{5}{4} = \frac{5}{4}\left(\frac{4}{5}r\right)$$

$$\frac{2}{3} \cdot \frac{5}{4} = \frac{5}{4}\left(\frac{4}{5}r\right) \leftrightarrow \frac{2}{3} \cdot \frac{5}{4} = \left(\frac{5}{4} \cdot \frac{4}{5}\right)r$$

$$\frac{2}{3} \cdot \frac{5}{4} = \left(\frac{5}{4} \cdot \frac{4}{5}\right)r \leftrightarrow \frac{2}{3} \cdot \frac{5}{4} = 1 \cdot r = r$$

(b) $$\frac{2}{3} \div \frac{4}{5} = \frac{\frac{2}{3}}{\frac{4}{5}} \cdot 1 = \frac{\frac{2}{3} \cdot \frac{5}{4}}{\frac{4}{5} \cdot \frac{5}{4}} = \frac{\frac{2}{3} \cdot \frac{5}{4}}{1} = \frac{2}{3} \cdot \frac{5}{4}$$

10 The student is having trouble here with the way that addition of fractions is defined where, with the usual introduction, one *must* get common denominators. One way of answering this student would be as follows: It would be true that one could get common denominators when multiplying fractions. For example,

$$\frac{1}{3} \cdot \frac{1}{2} = \frac{2}{6} \cdot \frac{3}{6} = \frac{6}{36} = \frac{1}{6}.$$

Notice, however, that there is no need to get a common denominator because of the way that multiplication of fractions is defined; i.e.,

$$\frac{a}{b} \cdot \frac{c}{d} = \frac{a \cdot c}{b \cdot d}.$$

Thus, in the above problem

$$\frac{1}{3} \cdot \frac{1}{2} = \frac{1 \cdot 1}{3 \cdot 2} = \frac{1}{6},$$

which is the same answer that we arrived at using common denominators.

11 The student seems to be using an incorrect algorithm for adding fractions. Most likely, the student is adding the numerators and/or denominators if these numbers are relatively prime (i.e., have greatest common divisor of 1); if the numerator and/ or denominator are multiples of each other, then the larger of these multiples is used. One approach that might be used with this student is to ask the student to add $\frac{2}{2} + \frac{4}{4}$. Since $\frac{2}{2} = 1$ and $\frac{4}{4} = 1$, then $\frac{2}{2} + \frac{4}{4}$ should equal 2. However, by the

algorithm that the student is using, $\frac{2}{2} + \frac{4}{4} = \frac{4}{4}$, which is certainly incorrect. Then, you might ask the student to add $\frac{2}{3} + \frac{3}{5}$. According to the algorithm that the student is using $\frac{2}{3} + \frac{3}{5} = \frac{5}{8}$. However, this algorithm is incorrect for adding rational numbers. See the article referenced in Section 1.5, Answer 3 for further information.

12 The teacher might respond as follows: "Yes, you would certainly have more pieces, but each piece that you would have would be smaller than each piece taken by the person who has 4/7 of the cake. In fact, five of your smaller pieces would be the same amount of cake as one of the pieces taken by the person who has 4/7 of the cake. Thus, since you have 20 of these small pieces, then you would have the same amount of cake as the person who took 4/7 of the cake." It would probably be beneficial to simulate the cutting of two cakes or similar objects into the appropriate sizes and showing the student the physical objects.

13 No, the student is not correct. Since the selling price equals the cost plus the profit, then selling price = 1/10 cost + cost; i.e., \$3300 = (1/10)c + c, and thus the cost is \$3000. Therefore, the profit is $(1/10) \cdot 3000 = \$300$.

14 What the teacher should have said was that a rational number is a number that can be expressed in the form of $\frac{p}{q}$, p, q integers, $q \neq 0$. Since $\frac{\sqrt{3} \cdot 3}{\sqrt{3} \cdot 4}$ can be expressed as $\frac{3}{4}$, then $\frac{\sqrt{3} \cdot 3}{\sqrt{3} \cdot 4}$ is a numeral that represents a rational number.

15 No, the student is not correct. There are $4\frac{4}{19}$ pieces each $4\frac{3}{4}$ feet long; i.e., there are 4 pieces and $\frac{4}{19}$ of a piece. Now, $\frac{4}{19}$ of a piece is $\frac{4}{19}$ of $4\frac{3}{4}$ feet, or 1 foot left over.

***Problem*** Find two different methods of showing the student that 1 foot is left over, not 4/19 of a foot.

## 1.6 Geometry

### (A) *Angles*

#### Questions

1 You are teaching your students the idea of the degree measure of an angle and you have related it to the circle. One student makes the assertion that the degree will vary because the angle intercepts arcs of different lengths depending upon the size of the circle. Is the student right? If so, what explanation will you give to get

around the difficulty presented? If the student is not right, then how will you resolve the student's misunderstanding?

2 A student asks the following question. "Since an angle is the union of two rays with a common endpoint and a triangle contains three angles, why doesn't the triangle look like this?"

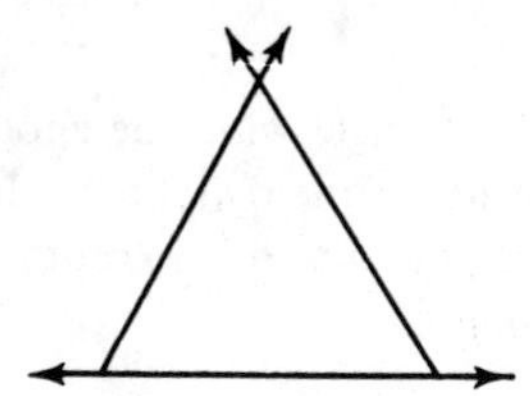

How do you respond?

3 How would you respond to the following question? Since two lines are perpendicular if they form a 90° angle with one another, is it necessary for them to be in the same plane to be perpendicular?

4 How would you respond if a student in your class asks, "Why can't you measure the size of an angle in inches?"

5 We are often faced with some very hard content questions to explain even at the seventh and eighth grade level. How would you help a student to answer the following question. "Why isn't a straight angle considered to be an angle in our definition?"

## Answers

1 One way of convincing the student that the degree measure of an angle will not vary is as follows: On a transparency construct a circle c with radius r. Using overlays with the same center as circle c, construct three or four other circles with different radii. On each circle have the student construct angles of various degree measures such as 90°, 45°, 30°, etc., taken one at a time. When all the overlays are put together, the student should see the following (Figure 1.6.1) for say 30°:

*FIGURE 1.6.1*

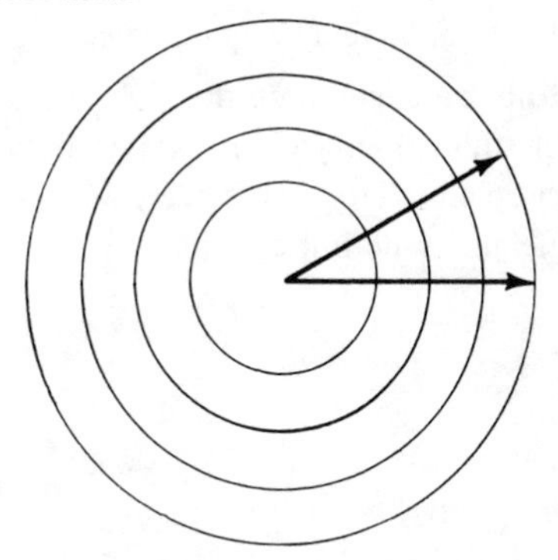

Thus, even though the angle intercepts arcs of different lengths in different circles, the angle being measured is still 30°.

2 The student is using imprecise language. The angles are *not contained* in the triangle, but the angles are *determined* by the triangle. Thus when we talk about the three angles of a triangle, we mean the three angles are determined by the three corners of the triangle.

3 Yes, it is. If two distinct lines $l_1$ and $l_2$ form a 90° angle with one another, then the lines intersect at the point, P, which is the vertex of the right angle. If we take any other point on $l_1$ except P and likewise any other point on $l_2$ except P, then we have the plane determined by these noncollinear points.

4 You might ask the student how this could be done. The student would probably suggest a method similar to the one shown in Figure 1.6.2. "Measure arc RS in inches as in the figure given, and use this number as the measure of angle ABC."

*FIGURE 1.6.2*

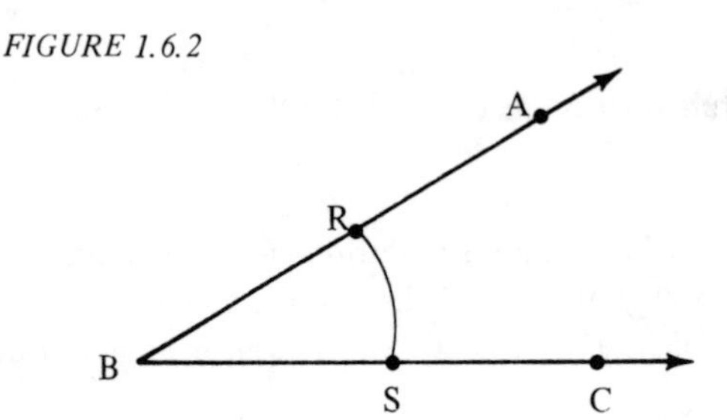


Of course, the basic problem here is that the measure of the angle will change depending upon how the points R and S are chosen. Since this is a highly unsatisfactory situation, measuring the size of angles in inches is not used. If one introduces the degree measure or radian measure at this level, one should certainly point out that these measures are independent of the "relative size" of the angle drawn.

5 Some texts do not consider a straight angle to be an angle because of the following limitations.

(a) A straight angle is impossible to distinguish from a line.

(b) A straight line angle has no unique vertex, sides, interior or exterior.

(c) These would be the only angles not contained in a unique plane.

When a straight angle is not considered to be an angle, then the idea of a linear pair is introduced to take care of the idea of supplementary angles; i.e., if two angles form a linear pair, then the angles are supplementary. However, whether a given text includes a straight angle as an angle in its definition of same is a matter of choice by the authors.

## (B) *Area*

### Questions

1 You ask your eighth grade students to find the area of the circle with a radius of five inches. One of your students then gives the following answer. "You cannot find the area of a circle, you can only find its perimeter." What is your response?

2 While you are teaching your seventh grade class the formulas for areas of simple polygons (triangle, square, etc.), a student excitedly says that she has a formula that does it all. She says that the formula is: $A_p = \frac{1}{2}h \cdot (t + b)$ where:

$A_p$ is the area of the polygon you want;

h is the length of the height of the polygon;

t is the length of the top segment;

b is the length of the bottom segment.

What do you tell the student?

3 A seventh grade student is trying to find the area of a rectangular lot 75 feet by 25 feet. The student reasons as follows: "I'll subtract 25 from 75 thus making it 50; then I'll add the 25 I subtracted to the other 25; then the problem reduces to 50 x 50 which is 2500. Therefore, the area of the lot is 2500 square feet." Is the student correct?

4 A student asks, "If two figures have the same area and the same perimeter, are the figures congruent?" What is your response?

### Answers

1 The student is correct. A circle is normally defined to be the set of points in a plane at the same distance from a fixed point called the center of the circle. Thus, a circle has no area but does have a perimeter called the circumference. The teacher meant to ask the students to find the area of the circular region with a radius of five inches.

2 A very good teaching strategy to use here is to first compliment the student for doing some excellent and creative thinking. Using positive reinforcement can be very beneficial to the student and the class. Then you might ask the student to demonstrate how the formula can be applied to the various polygons. Assuming that the student does this correctly for the simple polygons such as the triangle, square, rectangle, trapezoid, etc. you might ask her if the formula can be applied to figures such as the pentagon, hexagon, etc. You also might ask the student how the formula can be applied for the circle. She will have to do quite a bit of thinking in order to get the area of a circle, but it can be done.

However, for some polygons such as the regular hexagon, the student will get an incorrect answer, e.g.,the area of a regular hexagon is $\frac{1}{2} a \cdot p$, where a is the apothem of the hexagon and p is the perimeter. According to the student's formula, the area of the hexagon should be:

$$\frac{1}{2} \cdot 2a \cdot \left(\frac{p}{6} + \frac{p}{6}\right) = \frac{a \cdot p}{3}.$$

Thus, although the formula works for quite a number of polygons, it does not work for all polygons. Also consult:

Howard Fehr, *Secondary Mathematics: A Functional Approach for Teachers*, D. C. Heath & Co., Lexington, Ma., 1951, pp. 8-19.

***Problem*** Although a *direct* application of the student's formula does not work in all cases, the student has discovered a very important principle. Can you modify the student's formula so that it will work? If you have trouble consult the reference above.

3 For an answer to this question, read the following article by M. Walter.*

# A common misconception about area

MARION WALTER*
*Harvard University, Cambridge, Massachusetts*

*Marion Walter is an assistant professor at the Harvard University Graduate School of Education. Her concern is the education of teachers in the mathematics program. She conducts many in-service workshops and is particularly interested in the visual aspects of learning mathematics.*

**Can you help the bank manager?**

Recently a friend of mine had her living room floor scraped. The cost was to be reckoned by the square foot. She measured the floor to be 19½ feet long and 12½ feet wide. In order to know how much the job was going to cost, she had to know how many square feet were going to be scraped. Thus, she had to calculate the area of the floor. She remembered that one could find the area of a rectangle by "multiplying the length by the width." How many square feet: 19½ ft. × 12½ ft.? So that she wouldn't have to multiply numbers involving halves, she decided to calculate 19 × 13 instead of 19½ × 12½. (She subtracted ½ from the 19½ and added it to the 12½.) Draw yourself a diagram.

She realized that the calculation would have been still simpler if she had rounded off the numbers the other way; instead of changing the 19½ × 12½ to 19 × 13, she could multiply 20 × 12. (Multiplying by 20 is easier than multiplying by 19.) To check her calculation, she calculated both products. What a surprise she got when the two answers did not come out to be the

* Marion Walter is author of a booklet, *Boxes, Squares, and Other Things: A Teacher's Guide for a Unit in Informal Geometry*. The unit is available through the NATIONAL COUNCIL OF TEACHERS OF MATHEMATICS, 1201 Sixteenth Street NW, Washington, D.C. 20036.

*From *The Arithmetic Teacher*, April 1970, pp. 286-89. Reprinted by permission.

same! She "did it" several times. No, 19 × 13 = 247 and 20 × 12 = 240. She was stumped. "After all," she argued, "either way, I have taken off a half foot from one side and put it on the other—it should give the same answer."

The next day she took the vexing problem to her friendly bank manager who is usually only too pleased to explain any intricacy of income tax, and even complications on incorrect gas bills. He too was mystified. He calculated 20 × 12 and 19 × 13—no, the two products were *not* the same. One was 240 and the other, 247. He even calculated the exact answer (19½ × 12½ = 243¾) on his desk calculator. It did not worry him that the two calculations did not give the *right* answer—only that the two estimates did not give the *same* answer.

Your students may want to explore other floors—for example, a room 25½ × 20½ feet or 14½ × 8½ feet, or 17½ × 9½ feet, or 14¼ × 8¾ feet. Then ask your class to try to explain the dilemma of the bank manager. The drawings in figure 1, though not drawn to scale, may help.

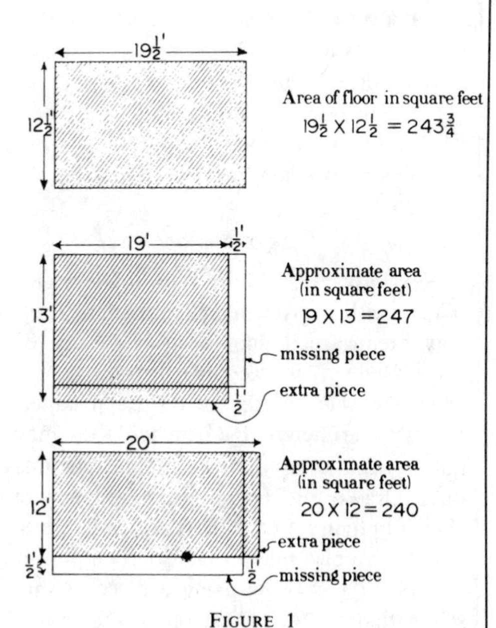


FIGURE 1

**Trouble with area and perimeter**

The opening story is a true one. Many children and adults have trouble dealing with areas and perimeters. They often think that two patterns with the *same* perimeter, though *different* in shape, must have the *same* area. As can be seen from figure 1, the perimeters of all three shapes are the same—they are 2 (19½ + 12½) feet, 2 (19 + 13) feet, and 2 (20 + 12) feet, so each is equal to 64 feet. My friend probably realized intuitively that she was not changing the perimeter when she "took off ½ foot from one side and added it to the other," and concluded therefore that the areas should remain the same too. It is interesting to note that the bank manager *was* prepared to have the two estimates differ from the real answer but not from each other.

Let us look for a moment at the relationship between the estimates and the correct answer. Notice that in our case one approximation is too small and the other too large. We have

$$240 < 243\tfrac{3}{4}$$

and

$$247 > 243\tfrac{3}{4}$$

which can be written 240 < 243¾ < 247. The bank manager may have noticed that the average of the two estimates

$$\left(\frac{240 + 247}{2}\right)$$

is 243½ and that the actual area (243¾) differs from the average of the two estimates by only ¼. Cutting a transparent paper model of each piece shown in figure 1 (but drawn to scale), and placing all 3 models on top of each other (figure 2) will help one to see the relationship.

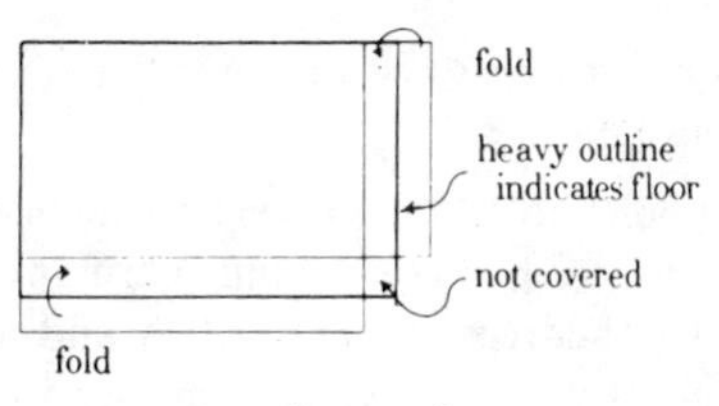


FIGURE 2

Fold the extra pieces over as shown to cover part of the model of the floor. The floor is now covered twice except for one small piece. The missing small piece has area ¼ square foot. Can you see how this is related to the difference between the correct answer and the average of the two estimates?

**Another example**

Before making some suggestions, let us consider another example where misconceptions often creep in.

FIGURE 3

When someone is asked to find the area of a pattern like that in figure 3, a common first reaction is, "I have not been taught a formula for finding the area of such an irregular shape!" Often, after some thought, such a person takes a piece of string to measure the perimeter of that shape. Then he uses that length of string to

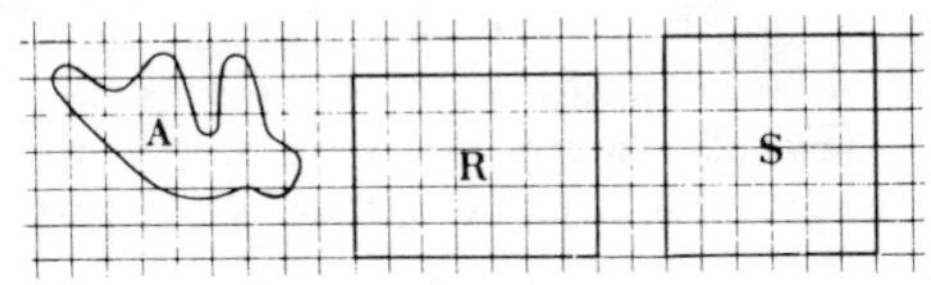


FIG. 4. The perimeter of each shape is 24 inches.

make a rectangle $R$ or square $S$ (see fig. 4). Then the area of such a rectangle (or square) is found by using the formula [$A = \ell \times w$], which he "knows." Great is the surprise when he finds out that the areas of $R$ and $S$ are not necessarily the same and that neither generally gives the area of $A$.

**What can be done to prevent such misconceptions?**

In order to try to prevent such misconceptions from taking hold permanently, first let children find perimeters and areas of figures without giving them any formulas. For example, let the children draw some irregular shapes as shown in figure 5.

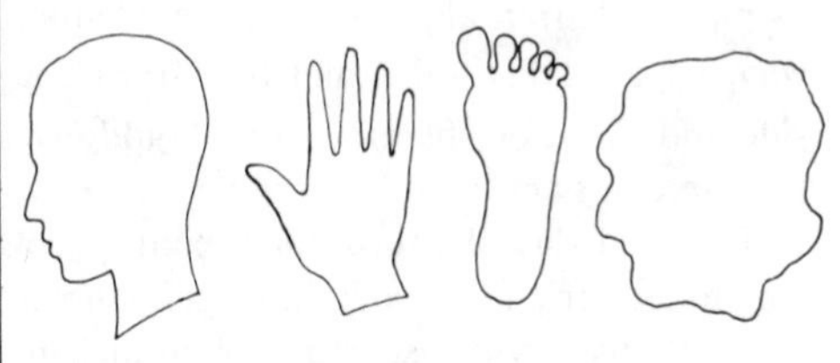

FIGURE 5

They can find the approximate perimeter of each by using string and laying it carefully on the figure. After straightening the string they can measure its length. You can also draw a large "park" on the floor or a playground and measure its perimeter by using a piece of string.

Work can be done with areas. Choose a small square as a unit. (For work preceding the use of square units see [2], for example.) Ask the children to find out approximately how many square units are needed to cover each of the shapes shown in figure 5. This can be done in a variety of ways:

> Children can cut out unit squares and lay them on the region and then estimate halves of squares to get a better approximation.
>
> Children can draw the shapes on graph paper of appropriate scale and count how many whole and how many half squares are inside the region.
>
> Can you think of another way?

Next ask the children to draw many shapes with a given fixed perimeter. They may create such shapes in a variety of ways. One way is to count edges of squares when the shape is drawn on graph paper. Using a perimeter of 24 inches, they may obtain numerous shapes. Some possible ones, drawn on 1-inch graph paper are shown in figure 4 ($R$ and $S$) and in figure 6.

One can also make shapes with a perimeter of 24 inches by using a piece of thin string that is 24 inches long (after being tied). One can trace any shape that the string makes. Some possible shapes are shown in figure 7.

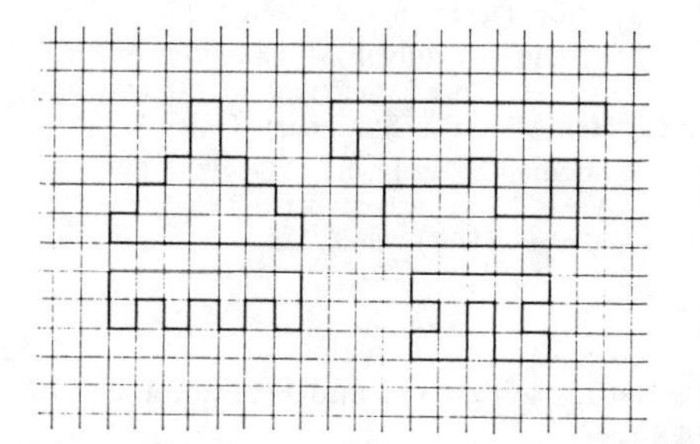

FIG. 6. The perimeter of each shape is 24 inches.

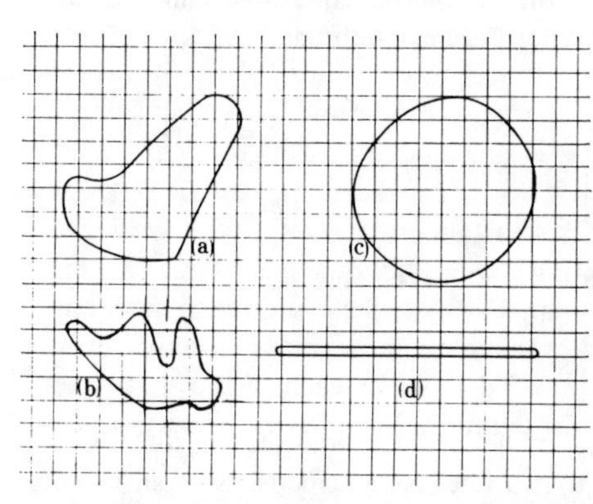


FIG. 7. The perimeter of each shape is 24 inches.

Many people will be surprised that with a given perimeter of 24 inches, shapes exist that have different areas. For example, we have made shapes with areas of 35 square inches, 36 square inches (see fig. 4), 17 square inches, 11 square inches, and 14 square inches (see fig. 6)—just to list a few. One can get a shape that has an area as small as one wants—for example, ½ square inch. (Can you describe such a shape?)

The areas of the shapes in figure 7 (each of which has a 24-inch perimeter) can be found approximately by counting squares and estimating half-squares. The shape in figure 7 that is roughly circular can be found to be approximately 45 square inches while shape 7*d* is approximately 4 square inches. What can you say about 7*b*? Can you make shapes with perimeter 24 inches that have an area of any number of square inches between 0 and 45? Figure 8 shows a shape whose perimeter is 24 inches and whose area is 34 square inches.

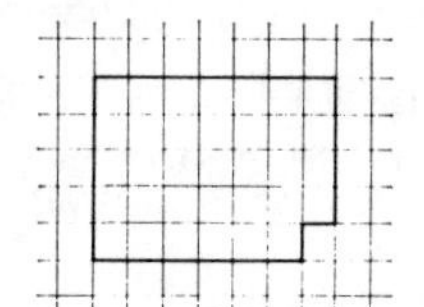

FIG. 8. 24-inch perimeter.

Investigate several such problems using different perimeters. What is the shape of the largest area you can get for a given fixed perimeter?

Turn the problem around now. Suppose you have a shape that has an area of 20 square inches, for example. Some such shapes are drawn on 1-inch graph paper in figure 9.

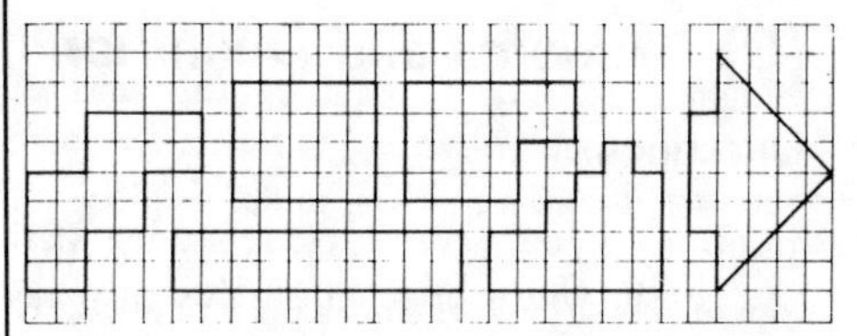

FIG. 9. The area of each shape is 20 square inches.

Find the perimeter of each shape, either by counting edges of squares or by using string. Are the perimeters of all these shapes the same? Can two different shapes with the same area have the same perimeter? Draw other shapes with an area of 48 square inches. Which has the smallest perimeter? Which has the largest? Is there a smallest? Is there a largest?

After several exercises like the ones described in this article, the children should no longer feel that shapes with the same perimeter always have the same area—or that shapes with the same area always have the same perimeter. You will find it useful to look at the references given below.

### References

1. Biggs, Edith, and MacLean, James R. *Freedom to Learn: An Active Learning Ap-*

*proach to Mathematics*. Ontario & Reading: Addison Wesley, 1969.
2. Bourne, H. N. "The Concept of Area." THE ARITHMETIC TEACHER 15 (March 1968): 233–43.
3. Holloway, G. E. T. *An Introduction to the Child's Conception of Geometry*. London: Rontledge & Kagan Paul, 1967.
4. Members of the Association of Teachers of Mathematics. *Notes on Mathematics in Primary Schools*. Cambridge: University Press, 1967.
5. Page, David. *Ways to Find How Many*. University of Illinois Arithmetic Project. Education Development Center, 1965.
6. The Schools Council. *Mathematics in Primary Schools*. 2d ed. Curriculum Bulletin No. 1. Her Majesty's Stationery Office, 1967.
7. Strangman, Kathryn. "Grids, Tiles, and Areas." THE ARITHMETIC TEACHER 15 (December 1968): 668–72.

4 For an answer to this question, read:

(1) Janet Jean Brougher, "Discovery Activities with Area and Perimeter," *The Arithmetic Teacher*, May 1973, pp. 382-85.

(2) Robert W. Prielipp, "Are triangles that have the same area and the same perimeter congruent?", *The Mathematics Teacher*, February 1974, p. 157.

# (C) *Axioms and Definitions*

## Questions

1 One of your seventh grade students asks, "You say that a point has no dimensions, i.e., in width or length. You also say that a line segment consists of an infinite number of points and that a line segment has a definite length. How can an infinite number of points each with no length considered separately, add up to a segment that has a definite length when considered all together. I don't understand." How can you help this student?

2 While beginning a unit on geometry with your junior high class, you make the statement that a point can be represented by a dot on the blackboard. You put the dot on the board and then erase it. One student then remarks that you have just erased the point. Is the student correct?

3 One student asks you the following question. "What is the difference between a ray and a half line?" How would you respond?

4 A student says, "Two half-lines make a line since ½ + ½ = 1." Is the student correct?

5 If a student asks the following question, "When working with rays such as $\overrightarrow{PQ}$, why is P called an endpoint rather than a beginning or starting point?" what response would you give?

6 A student is asked the definition of congruence and the reply is: "Congruence means equal." Is the student correct?

## Answers

1 This student is assuming that there is some kind of "infinite addition" of points. One must be extremely careful when dealing with infinite processes. Useful and intuitive properties of finite processes may not carry over. This idea becomes clearer once one has studied the calculus.

The student might even be shown that it is possible to consider a square as an infinite collection of lines (a square has a positive area, a line does not) and that a cube can be considered as an infinite collection of squares (a cube has a positive volume, a square does not).

2 No, the student is not correct. A dot is a physical representation of the abstract mathematical idea of a point just as a numeral is a written representation for a number. Since a point is an abstract idea, then it cannot be seen much less erased.

The main point for you to try to get across to the student is the difference between a geometric idea and the representation of that idea. The analogy of a number and its representation (a numeral) is a useful one to use in this situation. Another idea that might prove helpful is the distinction between a person and that person's name.

3 In many mathematics texts, there is no difference between the two. However, in some books such as those used in the elementary or middle schools, a half line is a ray $\overrightarrow{AB}$ with the endpoint or starting point "A" deleted. Thus one should look carefully at the definition in a particular book.

4 If the student is trying to get an intuitive grasp of the concepts involved, then it should be shown to the student that one can get a line from two half lines if the half lines include their endpoints as shown in Figure 1.6.3, but one could also get a figure such as shown in Figure 1.6.4.

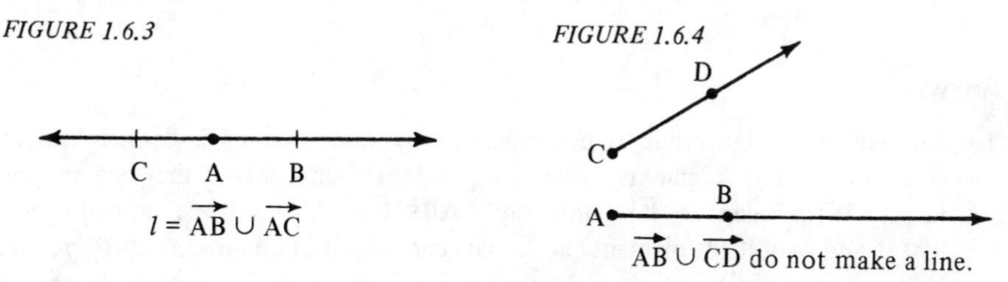


*FIGURE 1.6.3*

*FIGURE 1.6.4*

However, if a half line is defined as a ray $\overrightarrow{AB}$ with the endpoint A deleted, then the student is incorrect.

It should also be pointed out to the student that one cannot mathematically identify ½ with either a ray or a half line since this does not make any sense; e.g., what would ½ of a line mean?

5 In reality P should be called a beginning or starting point, but through traditional use the name endpoint has been commonly used. Perhaps the name came about because if one allows movement in either direction on this ray, then one of the *endpoints* would be P since no further movement along $\overrightarrow{PQ}$ would be allowed in that direction.

6 The student is incorrect. Equality is reserved for identity; e.g., two triangles are equal if and only if they are precisely the same set of points. On the other hand, at this level, congruence means that objects have the same size and shape. Thus, two triangles could be congruent and not equal.

## (D) *Measurement*

### Questions

1 One student makes the following comment. "Congruence and equal segments have the same relationship or meaning as equivalent and equal sets, right?" What is your response?

2 A student asks, "Why are there two different types of measures for weight, dry measure and liquid measure?" How do you reply?

3 An eighth grade student says: "36 in. = 1 yard, 9 in. = ¼ yard, $\sqrt{9 \text{ in.}} = \sqrt{1/4 \text{ yard}}$, 3 in. = ½ yard." Is this student correct?

4 If a student asks, "Why are there two different measurements for the ton?" how do you reply?

5 A student asks you if there is a difference between the following terms: 2 square feet and 2 feet square? What response do you give?

6 How would you answer if a seventh grade student asks you why the abbreviation *lb* is used for pound?

### Answers

1 Although the relationship is not precisely the same, there is a definite similarity. One can think of segments as being equivalent if and only if they are congruent; i.e., $\overline{AB}$ is equivalent to $\overline{CD}$ if and only if $\overline{AB} \cong \overline{CD}$. Also, sets are defined as equivalent if and only if the elements in the sets can be put in one-to-one correspondence. However, the notion of *equivalent* is used in two different senses and one cannot interchange the meaning. For example, one cannot use the word *equivalent* as used for sets in the meaning for segments because there are segments that can be put in one-to-one correspondence with each other and yet the segments are not congruent as shown:

A ——— B

C ————— D

These segments are equivalent in the set theoretic meaning of the word equivalent (one-to-one correspondence) but not equivalent as congruent segments.

However, equal segments mean the same as equal sets, i.e., the same set of points or the same set of elements.

2 Dry weight is a measure of weight while liquid weight is really a measure of volume. Liquid measure is especially useful in certain situations such as in pharmacies. Druggists often have bottles that measure say 6 fluid ounces, and they can put any liquid in these bottles and know that they thus have a certain volume. The liquid in these bottles is then equivalent in weight to 6 ounces of water. Dry measure on the other hand is measured by weight not volume. For example, cereal packages may not be filled to capacity (volume) due to settling, but you can be assured it weighs the required amount.

3 The student is incorrect. The problem here is one concerning dimensional analysis. The student could be corrected as follows:

$$36 \text{ in.} = 1 \text{ yard}$$
$$9 \text{ in.} = \tfrac{1}{4} \text{ yard}$$
$$\sqrt{9 \text{ in.}} = \sqrt{\tfrac{1}{4} \text{ yard}}$$
$$3\ \sqrt{\text{in.}} = \tfrac{1}{2}\sqrt{\text{yard}}$$

This is the point where the student in incorrect. One cannot disregard $\sqrt{\text{in.}}$ and $\sqrt{\text{yard}}$. One might continue the problem as follows:

$$3\sqrt{\text{in.}} = \tfrac{1}{2}\sqrt{36 \text{ in.}}$$
$$3\sqrt{\text{in.}} = \tfrac{1}{2} \cdot 6\sqrt{\text{in.}}$$
$$3\sqrt{\text{in.}} = 3\sqrt{\text{in.}}$$

Also consult the answer given in Section 2.13, Answer 6.

4 Our system of weights and measures was carried over into the colonies from Great Britain. These weights and measures were not derived from any scientific basis, and thus there was considerable lack of uniformity in the units used.

In essence there are really three different systems used for measuring the ton: the long ton, the short ton and the metric ton. The long ton is derived from Great Britain and was formulated as follows: One stone = 14 pounds, 8 stones = a hundredweight = 112 pounds, 20 hundredweights = a ton = 2240 pounds. Thus the *long ton*, or the ***net ton*** as it is sometimes called, is a measure left over from our relation with Great Britain. This measure is still used today in some foreign countries and is used in our country mainly in the shipping and mining industries.

The short ton is the one commonly used in our country. It is derived from our desire to use 100 pounds = a hundredweight and thus 20 hundredweights = 2000 pounds.

The metric ton is also used and is equal to 1.102 short tons.

5 Yes, there is a difference between the terms. When one writes 2 square feet, one is referrring to a unit of measuring area (1 square foot) and there are two of them as illustrated in Figure 1.6.5.

*FIGURE 1.6.5*

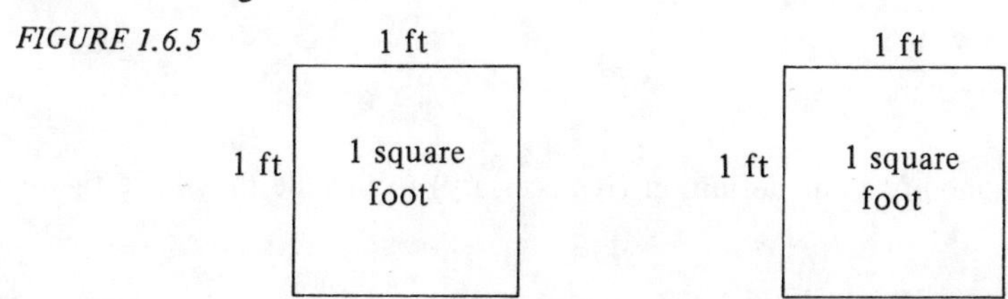

When one writes 2 feet square, one is referring to a square that has a length of 2 feet on each side as illustrated in Figure 1.6.6 and thus contains 4 square feet of area.

*FIGURE 1.6.6*

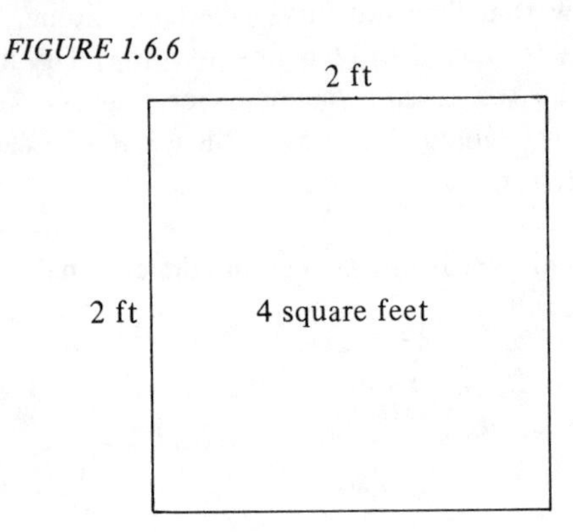


6 The abbreviation *lb* is derived from the Latin word *libra*, the ancient Roman word for pound.

## (E) *Polygons*

### Questions

1 You ask a student to draw a polygon on the board. One of your pupils draws this figure on the board. What is your reaction?

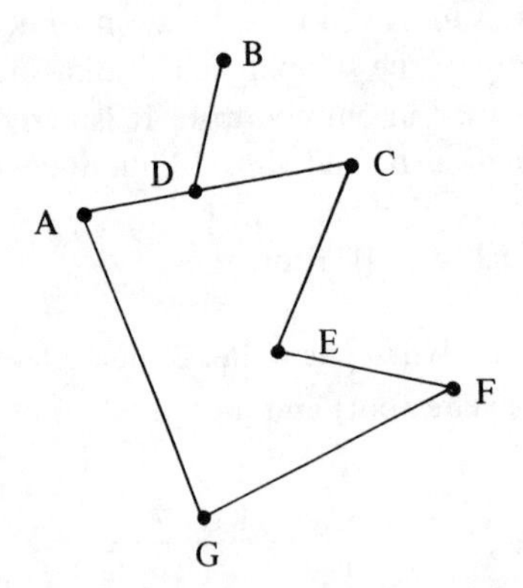


Could you modify your definition (if necessary) to include the above figure as a polygon?

2 A student in your class asks the following question. "What is the difference between a polygon and a polyhedron; is it that one just has sides and the other has area?" What is your response?

3 You state the following definition. "A polygon is the union of three or more coplanar segments, each of which intersects at least two of the other segments." After showing some examples of polygons, a student shows you the following figure and asks if it is a polygon. What do you tell this student?

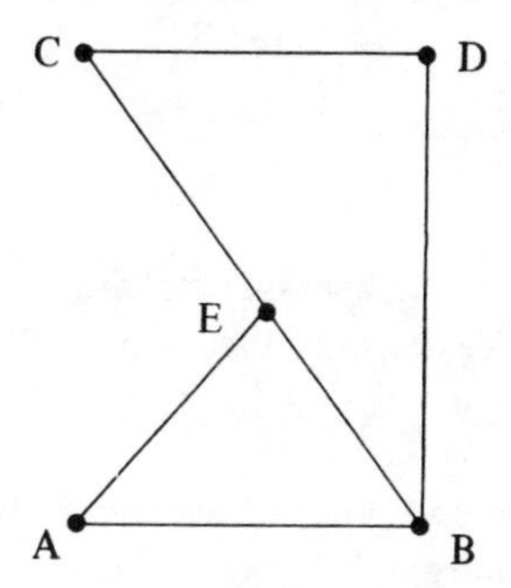


## Answers

1 Most likely this figure is one that we would not ordinarily want to classify as a polygon. Usually, at this level, we consider only simple, closed figures as polygons. A definition of a polygon that is usually given at this level is: a polygon is the union of three or more coplanar line segments joined at successive endpoints to form a closed figure, but otherwise the segments have no points in common and successive segments do not lie on the same line. If the following restrictions were all removed, then the figure shown could be classified as a polygon: (1) figure is closed; (2) segments contain no points in common other than endpoints; (3) successive endpoints are joined.

2 A polygon is a two-dimensional figure consisting of line segments, and the region determined by this figure has area. A polyhedron is a three-dimensional figure whose surface consists entirely of polygonal regions. Thus a polyhedron has both surface area (the sum of the areas of the polygonal regions) and volume. See the definition in Answer 1 of this section.

3 According to the definition made by the teacher, the figure is a polygon. Most likely, however, we would ordinarily not want the figure to be a polygon. Thus, it would seem to be the case that the teacher might want to modify the definition given by saying that each segment intersects exactly two of the other segments at the endpoints, but otherwise the segments have no other points in common.

## 1.7 Mean and Average

### Questions

1 A student in your eighth grade class is trying to solve the following problem. "A man drives a car for one mile at 60 miles per hour and for another mile at 120 miles per hour. What is his average speed?" The student reasons that the average speed should be

$$\frac{60 + 120}{2} = \frac{180}{2} = 90 \text{ mph.}$$

Is this student correct?

2 A student asks, "Is a baseball average (hitter's batting average) really an average or is it a probability since it is the number of successes over the number of trials?" How do you respond?

3 If a student asks, "Why do we need three different types of averages? Wouldn't one, like the mean, be good enough?" how do you reply?

4 Jack, a junior high student, is excited. "Miss Jones, I've found a new way to get the average of two numbers. Take 12 and 18, for example. Subtract 12 from 18; that's 6. Divide the difference by 2, you get 3. Add 3 to 12, and you get the average, 15." If you were Miss Jones, what would you tell Jack?

### Answers

1 No, the student is not correct. This error is one that many students and teachers make. The formula which the student is using assumes that the same amount of *time* is spent traveling at each speed. However, this assumption is violated in this problem; the cars travel at the same *distance* (one mile), but they do not travel this distance in the same amount of time. Traveling at 60 mph for 1 mile takes 1 minute while traveling at 120 mph for 1 miles takes ½ minute. The correct formula to use is the following:

$$\text{average speed} = \frac{\text{total distance}}{\text{total time}} \qquad (1)$$

Thus, in this problem the average speed would be $\frac{2}{1\frac{1}{2}} = \frac{4}{3}$ miles per minute or 80 miles per hour.

You might want to show the student that the formula he was using is a special case of (1), where the cars are traveling for the same time, as follows:

Let x = the amount of time that either car travels,
y = distance traveled at 60 mph,
z = distance traveled at 120 mph.

Thus,

$$\text{average speed} = \frac{y+z}{2x} = \frac{1}{2}\left(\frac{y+z}{x}\right) = \frac{1}{2}\left(\frac{y}{x} + \frac{z}{x}\right)$$
$$= \frac{1}{2}(60 + 120) = \frac{60 + 120}{2}.$$

2 A baseball average can be looked at either as an average or as a probability. For example, if we assign the numeral 0 for not getting a hit and the numeral 1 for getting a hit, then the arithmetic mean or average of the following 0 0 1 0 0 1 0 0 would be 2/8 or 1/4. Likewise a 0 can be deemed as not successful in getting a hit and a 1 as being successful in getting a hit. The above can then be interpreted as

$$\text{Probability of getting a hit} = \frac{2}{8} = \frac{1}{4}.$$

3 We will assume that the different kinds of "average" mentioned in this question refer to the three different kinds of measures of central tendency called the mean, median and mode. At a higher level one might be referring to the arithmetic, geometric and harmonic means, but these means are not usually studied in high school until an Algebra II course.

Each of the averages, mean, median and mode, has certain features that are useful to a person who would like a summary measure of a set of data. The choice of which of these three measures to use requires careful thought. For example, if four persons have incomes of $5000, $6000, $7000, and $8000, then one kind of average, the mean, would be $6500. If a fifth person, whose salary is $20,000, is added to the group, then the mean of this new group would be $9200. Now, if someone says that this last figure is the average salary in a company, would this measure be a good one to use in this situation? Obviously not, because it does not represent either group and is very misleading.

The problem here is that an extreme score affects the mean and can move the mean away from what would generally be considered the central area. In this case, the median would probably be a better measure to use. The main point here is that the choice of an average as a summary measure of a set of data depends on several factors. Two of these factors would be: (1) the shape of the distribution scores; (2) for what purpose will the summary measure be used. In either case, one of these averages might be better to use than another in order to get a useful interpretation of a set of data.

4 You might first commend Jack for devising this new method. Then you might ask him how he developed this process and if it works all the time or only with certain numbers. If it seems to work all the time, can he devise a justification or proof that it does. An algebraic proof, which shows that Jack is correct, is given in the following. Assume without loss of generality that $x \leqslant y$. Thus,

$$y - x = f, \quad f \geqslant 0 \quad \text{and} \quad x + \frac{f}{2} = x + \frac{y-x}{2} = \frac{2x + y - x}{2} = \frac{x+y}{2}.$$

## 1.8 Numbers, Operations, Properties and Algorithms

### Questions

1 How do you handle the following situation? A student says, "You can't divide 2 by 3 because 2 is less than 3." All reasonable efforts to convince the student that $2 \div 3 = 2/3$ fail.

2 A student asked to solve $\dfrac{7.2 \times 10^8}{1.2 \times 10^5}$ does the following.

$$(7.2 - 1.2) \times (10^{8-5}) = 6 \times 10^3.$$

Is the student correct?

3 A student asks, "Why is it when you subtract two negatives you have to change the sign of the subtrahend?" How would you respond?

4 Problem: $96 \div 6$.
Student Solution:

$$\left.\begin{array}{rl} 6 & \\ +6 & \text{two 6's} \\ \hline 12 & \\ +12 & \text{four 6's} \\ \hline 24 & \\ +24 & \text{eight 6's} \\ \hline 48 & \\ +48 & \text{sixteen 6's} \\ \hline 96 & \end{array}\right\} = \text{thirty 6's}$$

So the answer to $96 \div 6$ is 30. What is your response?

5 While you are discussing multiplication, one student asks you if it makes a difference if you say "2 times 3" or "2 multiplied by 3"? What is your response?

6 Tony can't understand why division by zero is undefined. He wants $0 \div 0 = 0$. His explanation is: "If you have nothing in one hand and you divide it by nothing, why don't you still end up with nothing?" How would you respond?

7 You give your seventh graders a short quiz to do in class. Add the following:

(a) $(2 + 3) + (-6)$ (b) $2 + (3 + -6)$ (c) $2 + -5$

(d) $-5 + 2$ (e) $(-2 + -3) + -4$ (f) $(-2 + -3) + 0$

One of the students hands in the following work. Answer: $-22$. Is the student correct?

8 The student (sixth grade, top section) was asked to divide 16 by 3 by annexing zeros. She asks, "How can you divide 16 by 3 after you annex zeros, when you can

not otherwise, if adding zeros is the same as adding nothing?" What is your response?

9 A seventh grade student makes the following statement. "It doesn't make any sense to me that when you divide a number like $-10$ into $-200$ you get 20 because that means when you divided a small number into an even smaller number you get a great big answer." How would you answer this student?

10 While doing the problem $6 \div 0$, a student reasons as follows: "$6 \div 0$ equals nothing. Since you are dividing by nothing, you get nothing as a result." How will you help this student?

11 You are trying to teach the concept of dividing by zero, and you have just explained that $0 \div 0$ is undefined because the value could have too many answers. One of your students then asks, "Why can't we just define $0 \div 0$ to be equal to one and always use it that way?" How would you respond?

12 A seventh grader does the following problem.

```
    5 4
3 |1512
   15
     12
     12
```

How would you help this student?

13 In your seventh grade class, a student has devised a new subtraction algorithm and shows you the following work.

```
   2        (1)  7 - 4 = 3
   3̸4            10 - 3 = 7
 - 17
   17       (2)  2(10) - 1(10) = 1(10)
```

Is the student correct?

14 While doing long division with your class, you observe that one of the students did this problem as follows:

```
     115
25 |625
    25
    37
    25
    125
```

How can you help this student?

15 A seventh grade student says that zero is nothing and wants to know if he is mathematically correct. What is your response?

16 An eighth grader asks you the following question. "We have said that if a and b are integers, then $a - b = a + (-b)$. Now, subtraction is not commutative, but the expression on the right above $(a + (-b))$ is commutative; I don't understand." How would you handle this situation?

17 While studying a unit on the properties of zero and one, a student says that not only is zero an additive identity but it is also an identity with respect to subtraction. Do you agree?

18 In your seventh grade class, you are correcting a set of homework exercises, and you come across one student who is doing the following:

(a) $3 + 0 = 3$ (b) $4 + 0 = 4$ (c) $-1/2 + 0 = -1/2$

(d) $0 + 1/2 = 0$ (e) $0 + (-2/3) = 0$ (f) $0 + 1/4 = 0$

What concept is the student having trouble with?

19 You are showing a seventh grade student the standard method of dividing the following example.

```
      13
25 ) 3420
     25
      92
      75
     170
```

You get to this point and tell the student that 25 does not go into 1 so look at 17; the student says that it does, since the 1 represents 100 and 25 goes into 100, the answer being 4 times. How do you respond to his answer?

20 While working on a unit covering negative integers, you make the statement that $-3 < -1$. However, a student raises his hand and says that he does not believe it. He says that if you have a debt of three dollars then that is larger than a debt of one dollar. How do you respond to this student and how would you convince him that $-3 < -1$?

## Answers

1 One approach to handling this situation might be to ask the student what $2 \div 3$ means in terms of multiplication. Hopefully the student would respond that division is defined in terms of multiplication so that $2 \div 3 = c$ if and only if $2 = 3 \cdot c$. Then you would ask the student what number c would satisfy the latter equation. By direct substitution, the fraction 2/3 is a solution of $2 = 3 \cdot c$.

Another approach to the question might be an intuitive one using geometry. You could show the student how to divide the rectangle shown in Figure 1.8.1 into three congruent parts. Since this rectangle can be thought of as two squares put together, the student could be shown Figure 1.8.2 where the rectangle is divided into three congruent parts (one of the parts is shaded).

*FIGURE 1.8.1*

*FIGURE 1.8.2*

Since the area of the whole rectangle is two units, the area of one of the congruent parts in Figure 1.8.2 (such as the shaded part) must be 2/3 units. This is easily seen in Figure 1.8.2 because the shaded part in each square is 1/3 unit and there are two shaded parts.

2 In this particular case, the student is correct. However, in general, the method that the student used is incorrect. For example,

$$\frac{12 \times 10^4}{2 \times 10^2} = \frac{12}{2} \times \frac{10^4}{10^2} = 6 \cdot 10^2,$$

not $(12 - 2) \times (10^{4-2}) = 10^3$.

***Problem*** Under what conditions will the student's method work?

3 According to the definition of subtraction for integers, a – b = a + (–b). Thus, if b is a negative number, (–b), the additive inverse of b, is a positive number.

Another approach to this question might be to ask the student to graph a problem such as –2 – (–3) on the number line. If the student falters, ask the student to graph (–2) + (–3) on the number line, which is illustrated as follows:

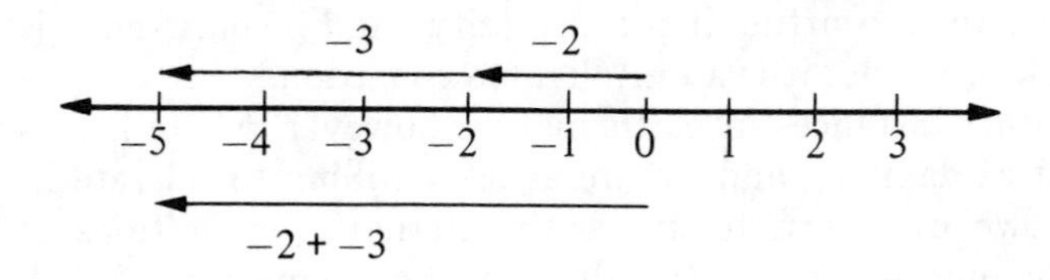


Then tell the student that since subtraction is the inverse operation of addition, the arrow representing –(–3) must go in the opposite direction as illustrated:

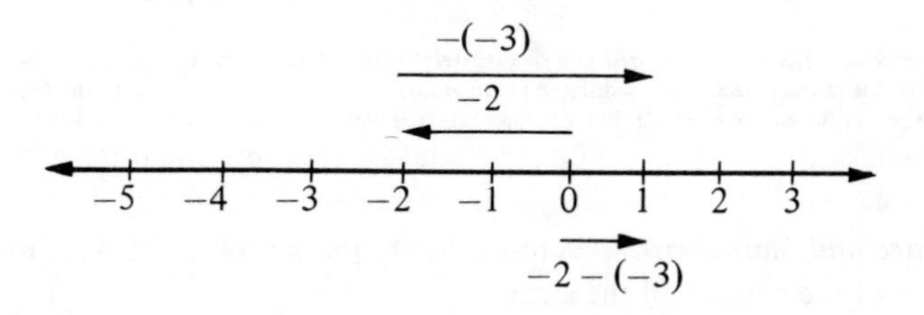

Since this is exactly how $-2 + 3$ would be represented, it must be the case that $-2 - (-3) = -2 + 3$.

4 Ask the student to multiply 30 by 6. The student should realize that if $96 \div 6 = 30$ then $30 \cdot 6 = 96$. Assuming that the student multiplies correctly, he should realize that something is incorrect. Then you might point out that the mistake made was adding together all of the 6s, in which case, some 6s have been added more than once. The correct answer is simply the number of 6s recorded in the last line. A possible strategy to use with this student is to ask him to do the following problem: $24 \div 6$. The student should realize that $24 \div 6 = 4$, and he would get the answer using the algorithm that he developed as follows:

$$\left.\begin{array}{r} 6 \\ +6 \end{array}\right\} \text{two 6s}$$
$$\left.\begin{array}{r} 12 \\ +12 \end{array}\right\} \text{four 6s}$$
$$24$$

Notice that the correct answer is the number of 6s recorded in the last line, not the total number of 6s (6 of them) recorded.

5 For an answer to this question, read the article by J. Weaver.*

## A Nontrivial Issue of Symbolism in Multiplication

J. Fred Weaver

*The University of Wisconsin, Madison, Wisconsin*

Consistency is not necessarily a virtue in all things, and inconsistency in some things does not necessarily imply lack of virtue. But in the case of a deductive mathematical system, for instance, consistency is more than a virtue; it is a necessity. And inconsistency is more than lack of virtue; it is an intolerable condition.

There are other instances in mathematics, however, where inconsistency is not as damning and we are at least willing to tolerate it. For example, we can learn to live with alternative definitions of *trapezoid*,[1] providing we understand the logical consequences of each and are aware of which definition is intended within a given context.

But we should pause and reflect upon some of these tolerated inconsistencies within the framework of mathematical learning among young children at the elementary-school level. It is not unlikely that

[1] Under one definition a trapezoid is characterized as a quadrilateral with *at least* one pair of parallel sides, under another definition it is characterized as a quadrilateral with *exactly* one pair of parallel sides. In the first case, *parallelogram* is a proper subclass of *trapezoid;* in the second case, parallelogram and trapezoid are disjoint classes.

*From *School Science and Mathematics*, October 1968, pp. 629-30, 639-42. Reprinted by permission of the editor and the author.

certain inconsistencies inhibit pupils' mathematical understanding. A case in point well may be the way in which we commonly interpret the symbol for the operation of multiplication [$\times$], in contrast with the way in which we interpret the operational symbols for addition [$+$], subtraction [$-$], and division [$\div$].

### An Interpretation of $+$, $-$, $\div$

Let us first consider the following illustrations:

(1) $12 + 4 = 16$ [12 *plus* 4 is equal to 16]

(2) $12 - 4 = 8$ [12 *minus* 4 is equal to 8]

(3) $12 \div 4 = 3$ [12 *divided by* 4 is equal to 3]

Virtually without exception, mathematics program for elementary-school children reflect a similar interpretation for each of these three instances: the first number, 12, is treated as the *operand* [the number operated on] and the second number, 4, is treated as the *operator* [which "acts" on the first number in accord with the operation specified ($+$ or $-$ or $\div$)]. The terms *operand* and *operator* are not used with elementary-school pupils, to be sure; but the *ideas* named by these technical terms clearly are in evidence.

### Interpretations of $\times$

Now let us consider this illustration:

(4) $12 \times 4 = 48$

If we read "12$\times$4" as "12 *multiplied by* 4," we are consistent with the preceding interpretation for addition, subtraction, and division: the first number, 12, is treated as the operand and the second number, 4, is treated as the operator [in this instance in accord with the operation of multiplication].

But suppose we follow a much more common practice at the elementary-school level and read "12$\times$4" as "12 *times* 4." Then we are treating the first number, 12, as the *operator* and we are treating the second number, 4, as the operand.

These two interpretations clearly do *not* mean the same thing. For example, some persons suggest that we associate the multiplication of counting numbers with the union of equivalent disjoint sets. If this is the case, "12 *multiplied by* 4" is associated with equivalent sets of 12 things, and the product of 12 and 4 is the number of elements in the union of 4 such sets. On the other hand, "12 *times* 4" is associated with equivalent sets of 4 things, and the product of 12 and 4 is the number of elements in the union of 12 such sets. [In either instance, of course, the equivalent sets are disjoint.]

Material on multiplication in most professional books, articles, etc. for elementary-school teachers usually includes no mention of this ambiguous interpretation of the operational symbol, $\times$. A notable exception, however, is found in Fehr and Phillips' opening discussion of "Multiplication of Cardinal Numbers" in which the following point of view is expressed:

> "It might be well if the 'times' language were never introduced, since it makes the symbols for a multiplication the only ones children have to 'read backward.' Some of the new textbooks are not using it for this reason. However, it does have certain advantages, and it is widely used. When teachers use the language 'times' and 'multiplied by,' they should know which numeral refers to the operator and which to the operand and express the operation accordingly in any work they do with a class. Since multiplication is commutative, it is not a serious matter if a child gets an occasional example 'turned around.' "[2]

### Looking Further

But let us look a bit further. In particular, let us look to certain mathematical sentences which relate the operations of multiplication and division. We shall do this, however, by first considering some mathematical sentences which relate the operations of addition and subtraction. The following properties are true for every real number $a$ and every real number $b$:[3]

(5) $(a + b) - b = a$ [e.g., $(12 + 4) - 4 = 12$]

(6) $(a - b) + b = a$ [e.g., $(12 - 4) + 4 = 12$]

(7) $a - b = a + {}^{-}b$ [e.g., $12 - 4 = 12 + {}^{-}4$]

In each of these cases it makes sense mathematically to treat $a$ [or 12] as the operand and to treat $b$ [or 4] as the operator at the outset. The remaining part of each sentence is consistent with this initial interpretation of operand and operator.

Now let us turn to the counterparts of properties (5), (6), and (7) for the operations of multiplication and division, where $a$ is any real number and $b$ is a real number such that $b \neq 0$:[4]

(8) $(a \times b) \div b = a$ [e.g., $(12 \times 4) \div 4 = 12$]

(9) $(a \div b) \times b = a$ [e.g., $(12 \div 4) \times 4 = 12$]

(10) $a \div b = a \times \frac{1}{b}$ $\left[\text{e.g., } 12 \div 4 = 12 \times \frac{1}{4}\right]$

The sense of each of these properties casts $b$ [or 4], and not $a$, in the role of an *operator*, with the consequent interpretation of "×" as "multiplied by" rather than "times."

Notice that if we read the left-hand side of sentence (8) as "($a$

[2] Howard F. Fehr and Jo McKeeby Phillips, *Teaching Modern Mathematics in the Elementary School.* (Addison-Wesley, 1967). pp. 101–102.

[3] These properties also are valid for some but not all distinguished subsets of the real numbers. For instance, they continue to be valid if $a$ and $b$ are rational numbers, or if $a$ and $b$ are integers. But suppose that $a$ and $b$ are whole numbers $[W = \{0, 1, 2, 3, 4, 5, 6, \cdots\}]$. Sentence (5) remains valid for every $a \in W$ and every $b \in W$. Sentence (6), however, is meaningful and valid only when $a \geqq b$. And sentence (7) is meaningless, since the concept of an additive inverse does not even apply within the set of whole numbers [i.e., for $b \in W$, ${}^{-}b \notin W$]. Notice also that sentence (6) is a logical consequence of the mathematical definition of subtraction in terms of addition: $a - b = c$ is true if and only if there exists a number $c$ such that $c + b = a$. Also, sentence (5) is a logical consequence of the implication: if $a + b = c$, then $c - b = a$.

[4] Observations analogous to those in footnote 2 apply here but are not stated explicitly. The reader may wish to do so for himself, however.

*times b) divided by b*," then the intended sense of this "doing-undoing" property is *not* maintained. Similarly, the intended sense of (9) and (10) is *not* maintained if "×" is read as "times" [and "÷" is read as "divided by."][5]

If we insist upon reading "×" as "*times*" and still wish to maintain the intended sense of the property exemplified by sentence (8), *without involving the additional property of commutativity of multiplication*, we have two alternatives. If $b$ is to be an operator, then we should write

(8.1) $(b \times a) \div b = a$ [e.g., $(4 \times 12) \div 4 = 12$]

On the other hand, if $a$ is to be an operator, then we should write

(8.2) $(a \times b) \div a = b$ [e.g., $(12 \times 4) \div 12 = 4$],

in which case we must apply the restriction that $a \neq 0$ rather than $b \neq 0$.

It should not be necessary to illustrate specifically the fact that we encounter similar problems with sentences (9) and (10) if we read "×" as "*times*" and wish to maintain the intended sense of the property involved in each case.

### Arrays

In an earlier article in this journal I advocated a basic characterization of the multiplication operation in terms of arrays.[6] In that article I was concerned principally with some other issues and, to avoid beclouding them, more or less followed the common practice of treating $a$ as the *operator* in expressions of the form $a \times b$ [read as "*a times*" $b$]. Now, however, I wish to emphasize the desirability of treating $b$ as the operator in connection with an array characterization of multiplication.

We commonly name an array with 5 rows and 7 columns, for instance, as a 5-*by*-7 array [Fig. 1]. This 5-by-7 array, and every 5-by-7 array has 5×7 or 35 elements.

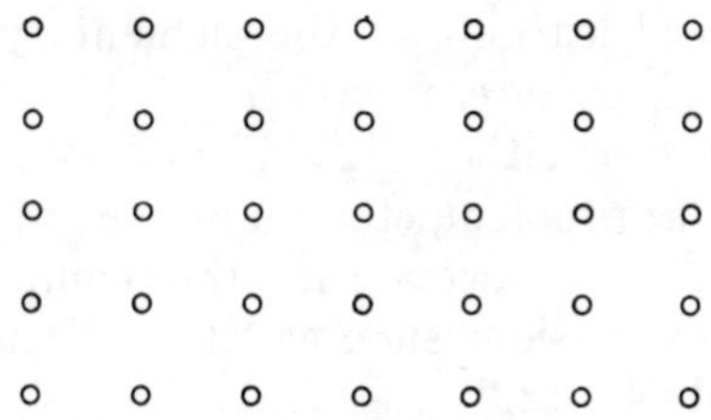

o o o o o o o
o o o o o o o
o o o o o o o
o o o o o o o
o o o o o o o

Fig. 1. A 5-by-7 array.

[5] At some mathematical levels we prefer to symbolize "$a$ divided by $b$" as $\frac{a}{b}$ or as $a/b$ rather than as $a \div b$. In each of these instances the operator is $b$, so that $a \div b = b = a/b$. These symbols,—particularly the last one,—should not be confused with the symbol $a|b$, which we may read as "$a$ divides $b$" [i.e., $a$ is a factor of $b$]. In the case of $a|b$, the operator is $a$ rather than $b$.

[6] J. Fred Weaver, "Multiplication Within the Set of Counting Numbers." School Science and Mathematics, 67: 252–70; March 1967.

More generally, an array with $a$ rows and $b$ columns is termed an $a$-*by*-$b$ array, and has $a \times b$ elements [where $a$ and $b$ are counting numbers.][7]

When we speak of a 5-*by*-7 array, for instance, it seems sensible and consistent to think of $5 \times 7$ as "5 *multiplied by* 7." In general, when we speak of an $a$-*by*-$b$ array, it seems sensible and consistent to interpret $a \times b$ as "*a multiplied by b*."

Notice also that when a 5-by-7 array, for instance, is viewed as a union of disjoint equivalent sets, it is preferable to think of the array as being generated in the form of Figure 2 rather than in the form of Figure 3 if we are to interpret "$5 \times 7$" as "5 *multiplied by* 7."

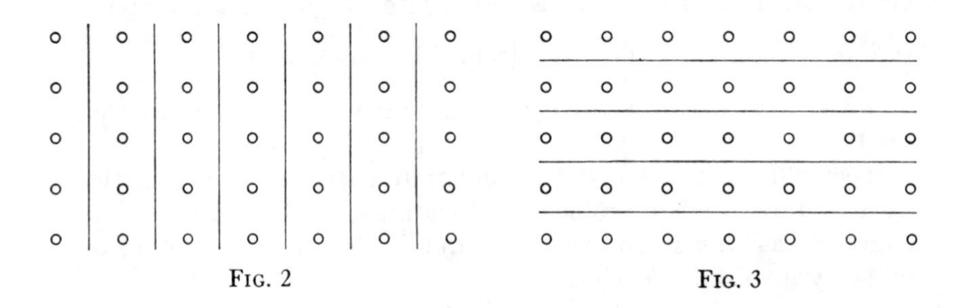

FIG. 2 FIG. 3

The sense of Figure 3 would lead to interpreting "$5 \times 7$" as "5 *times* 7" rather than as "5 multiplied by 7," if we are thinking of the array as being generated in relation to the union of disjoint equivalent sets.

## IN CONCLUSION

It is true that the symbol for multiplication [$\times$] may be interpreted ambiguously without affecting the product of two real numbers. Since multiplication is commutative [when applied to real numbers], "5 *times* 7" and "5 *multiplied by* 7," for instance, indicate the same product, 35. In general, for every real number $a$ and every real number $b$, "$a$ times $b$" and "$a$ multiplied by $b$" designate the same real number.

At certain stages of mathematical learning this ambiguity undoubtedly is of trivial consequence. But this is *not* a trivial matter for children's mathematical learning at the elementary-school level. If we tolerate an ambiguous interpretation of "$12 \times 4$," for instance; if we are willing to accept either "$12 \times 4$" or "$4 \times 12$" as expressing the sense of a particular problem, etc.; and if we are just as free with "$12+4$" and "$4+12$,"—then we are inviting confusion on the part of children between expressions such as "$12-4$" and "$4-12$," and between "$12 \div 4$" and "$4 \div 12$."

Let us attempt to build for elementary-school pupils an unambiguous interpretation of the symbol, $\times$, for multiplication. And in the interests of advantages likely to be gained by consistency of interpretation in relation to the symbols for addition [+], subtraction

[7] As indicated in the preceding article, the array interpretation of $a \times b$ for counting numbers may be extended to whole numbers, to cover those instances in which $a=0$ or $b=0$, and $a \times b=0$.

[−], and division [÷], let us consider seriously the interpretation of "×" as "*multiplied by*" rather than as "times." This non-trivial issue at the elementary-school level deserves more than token attention in terms of empirical exploration and investigation.

*A Postscript*

The manuscript for this article was prepared and submitted before I read in the November 1967 issue of SCHOOL SCIENCE AND MATHEMATICS Professor Rappaport's reaction to my previous discussion of "Multiplication Within the Set of Counting Numbers,"[8] in which he contends that I emphasized "logic at the expense of psychology."[9] This reaction has given me no cause to modify either the suggestion I advance in this present article or the position I advocated in my previous discussion.

I am intrigued by the fact that my emphasis upon "logical" [mathematical] considerations which Rappaport interprets to be at the "expense of psychology" can be viewed, instead, to have *support* from "psychology." For instance: In a recent discussion of meaningful verbal learning, Ausubel[10] states: "It is self-evident that before new ideas can be meaningfully learned, they must be adequately discriminable from similar established ideas in cognitive structure. If the learner cannot discriminate clearly, for example, between new idea $A'$ and previously learned idea $A$, then $A'$ enjoys relatively little status as a separately identifiable meaning in its own right, even at the very onset of its incorporation into cognitive structure."[11]

Rappaport asks, "Is it not possible to develop mathematical principles in accordance with sound psychological principles?"[12] Certainly it is,—and I am very much interested in doing so. In view of the principle mentioned by Ausubel, I have both *psychological and mathematical* reasons to question Rappaport's contention that "Repeated addition serves as an excellent introduction to multiplication,"[13] and to favor an array interpretation independent of addition.

Finally, interested readers may wish to relate two articles from another journal to the issue in question:

Robert D. Bechtel and Lyle J. Dixon, "Multiplication—Repeated Addition?" *Arithmetic Teacher*, 14: 373–76, May 1967.

David Rappaport, "Multiplication—Logical or Pedagogical?" *Arithmetic Teacher*, 15: 158–60; February 1968.

---

[8] David Rappaport, "Logic, Psychology and the New Mathematics." SCHOOL SCIENCE AND MATHEMATICS, 67: 681–85; November 1967.

[9] *Ibid.*, p. 684.

[10] David P. Ausubel, "Facilitating Meaningful Verbal Learning in the Classroom." *Arithmetic Teacher*, 15: 126–32, February 1968.

[11] *Ibid.*, pp. 131–32.

[12] Rappaport, *ibid.*, pp. 681–82.

[13] *Ibid.*, p. 684.

6 One way of solving this problem is: Go back to the definition of "÷" where $a \div b = c$ if and only if $a = bc$. If $a = 0$ and $b = 0$, we have

$$0 \div 0 = c$$

if and only if

$$0 = 0 \cdot c. \tag{1}$$

*Any* number c will make (1) a true statement. When this situation develops, it is customary for mathematicians to say that the particular operation in question is undefined (or undetermined).

An alternate approach to the given problem is as follows. It is frequently the case that an expression such as $6 \div 2$ is intuitively presented as the number of times two objects can be subtracted from six objects before one encounters a negative number. In the case of $0 \div 0$, one would be asking, "How many times can one subtract zero objects from zero objects before one encounters a negative number?" This could be done once, twice, etc. Again, in this case, there is not a unique answer.

7 The problem is with the teacher and not the student. The question is very ambiguous, although the teacher probably wanted the answer to each individual part. What the student did was to find the answers to each individual part and add them together. Since the answer given by the student is correct under this interpretation, it would be unjust to mark the answer incorrect.

8 The student is not correct in saying that we cannot divide 16 by 3 without annexing zeros. This can be accomplished by two different methods as follows:

(1) Since $16 \div 3 = \frac{16}{3}$, then $16 \div 3 = 5\frac{1}{3} = 5 + \frac{1}{3}$.

$$\frac{1}{3} = \frac{1}{10} \cdot \frac{10}{3} = \frac{1}{10}\left(3 + \frac{1}{3}\right) = \frac{3}{10} + \frac{1}{10} \cdot \frac{1}{3}$$

Thus $16 \div 3 = 5 + \frac{3}{10} + \frac{1}{10} \cdot \frac{1}{3}$.

$$\frac{1}{10} \cdot \frac{1}{3} = \frac{1}{100} \cdot \frac{10}{3} = \frac{1}{100}\left(3 + \frac{1}{3}\right) = \frac{3}{100} + \frac{1}{100} \cdot \frac{1}{3}$$

Thus $16 \div 3 = 5 + \frac{3}{10} + \frac{3}{100} + \frac{1}{100} \cdot \frac{1}{3}$ which can be written as $5.33\frac{1}{3}$. It is evident that this process can be continued as far as one would like to go.

(2) Since $16 \div 3 = \frac{16}{3}$, then $16 \div 3 = \frac{16 \cdot 100/3}{100}$. We can find $\frac{1600}{3}$ by using the division algorithm for natural numbers.

```
   533
3 |1600
   15
    10
     9
    10
     9
     1
```

Thus $\frac{1600}{3} = 533\frac{1}{3}$, and therefore $\frac{1600/3}{100} = 5.33\frac{1}{3}$.

In this situation, we can see that in order to get a decimal fraction we first move the decimal point in the numerator two places to the right (by multiplying by 100). Then we move the decimal point to the left two places by dividing by 100. The net result is to leave the decimal point where it was in the first place and supply zeros as needed. Then we carry through the division algorithm as if there were no decimal point present. The act of annexing zeros is thus simply the result of devising a shorter algorithm that accomplishes the same thing as the longer processes presented.

9 There are several ways of explaining this result to a student as follows:

(1) Suppose that you owe someone \$200 represented by –200, and you are going to repay the loan by giving the person \$10 a day, represented by –10. Then, how many days will it take you to pay the debt? The main idea is to try to get the student to see division as repeated subtraction.

(2) Division can be defined in terms of multiplication as follows: $a \div b = c, b \neq 0$, if and only if $a = b \cdot c$. Thus $-200 \div -10 = c$ if and only if $-200 = -10 \cdot c$; i.e., $c = 20$.

(3) One could also show the result on a number line.

***Problem*** Show $-200 \div -20$ on a number line.

10 One way of doing this problem is to go back to the definition of "÷". $6 \div 0 = b$ if and only if $6 = 0 \cdot b$. However, $0 \cdot b = 0$ and $6 \neq 0$. Thus, there is no number b such that $6 \div 0 = b$.

An intuitive way of looking at this problem is as follows: $6 \div 0$ can be thought of as repeated subtraction. Thus $6 \div 0$ asks the question, "How many times can you subtract zero from 6 until the six objects are gone." You can never do this, and thus $6 \div 0$ is not defined.

An alternate way of looking at the problem is to first look at related problems. For example, $6 \div 2$ asks the question, "How many times can you partition 6 objects into groups of 2." Thus $6 \div 2 = 3$. $6 \div 1$ likewise would be 6. Then $6 \div 0$ would ask, "How many times can you partition 6 objects into groups of 0." It can't be done.

11 If $0 \div 0$ were defined to be equal to one, then one would be led to make the following deduction. Since the fraction $a/b = a \div b$, then the fraction a/b would be equivalent to 0/0, which is equal to 1. That is, every fraction would be equivalent to a fraction, which is a name for 1, and thus we would have just one equivalence class of fractions all of which are names for the number 1. Since the idea of equivalence classes of fractions is such an important idea when working with fractions, then it follows that defining $0 \div 0$ to be equal to 1 is not justified.

12 One way of helping this student is as follows: Ask the student to multiply 54 by 3. Assuming that the student does this correctly, it should be obvious that something is wrong. The following procedure may then be shown to the student.

```
      4 }
    500 }→504
3 )1512
   1500
     12
     12
```

By employing this procedure, the student should realize that 5 in the quotient refers to 500.

13 The student is correct. Intuitively, in the first step, the student is finding out how many objects are left over in the units place in the subtrahend $(7 - 4)$. Then the number 3 is subtracted from the minuhend in the form of $10 - 3$. Since the student has used up one ten in the first step, then, in the second step, the student must reduce the number of tens by one and then subtract the number of tens. In general, let the subtraction be represented by

$$\begin{array}{r} ab \\ -\underline{cd} \end{array} \quad \text{where } b < d.$$

Then the student would do the following:

$$\begin{array}{r} ab \\ -\underline{cd} \end{array} \qquad \begin{array}{ll} (1) & d - b = x \\ & 10 - (d - b) = (10 + b - d) \\ (2) & (a - 1) \cdot 10 - (c \cdot 10) = (a - 1 - c) \cdot 10 \end{array}$$

14 This student has a problem with place value. Thus, one way of helping him would be to rewrite the way he has done the problem as follows:

$$\begin{array}{r} 5 \\ 10 \\ \underline{100} \\ 25\overline{)625} \\ \underline{250} \\ 370 \\ \underline{250} \\ 125 \\ \underline{125} \end{array}$$

Then, ask the student if $25 \cdot 100 = 250$. The student should then see that the first "1" is placed in the wrong place. If the student wants to do the problem in the manner illustrated, then the following should help.

$$\begin{array}{r} 5 \\ 10 \\ \underline{10} \\ 25\overline{)625} \\ \underline{250} \\ 375 \\ \underline{250} \\ 125 \\ \underline{125} \end{array}$$

15 Mathematically, the student is not correct. For example, if one knows that it is zero degrees centigrade outside does this mean that there is no temperature? Likewise, in bowling, if a zero is placed on the scoreboard, then this would indicate that the person had a turn. If nothing is put on the scoreboard, then the person has not yet taken his turn. The student is trying to equate the empty set with the set containing zero. Mathematically, the two concepts are different although in normal everyday use we usually use the terms interchangeably with no resulting ambiguity.

16 The student is assuming that since addition is commutative, it should also be true that subtraction is commutative; i.e., if $a + (-b) = (-b) + a$, then $a - b = a + (-b)$. However, $b - a = b + (-a)$, and this is the point where the student is confused. The student thinks that $b - a$ should equal $-b + a$. Thus, the fact that addition is commutative just happens to be a property of addition that is an added bonus and has nothing to do with $a - b = a + (-b)$.

17 No, the student is not correct. For example, $2 - 0 = 2$, but $0 - 2 = -2$, not 2. Thus, 0 is a right-handed identity but not a left-handed identity with respect to subtraction. Remind students that if $a \in S$ and # is the binary operation in S, then i is an identity with respect to S and # if and only if $a \# i = i \# a = a$ for all $a \in S$.

18 The student does not seem to understand that zero is the additive identity; i.e., $a + 0 = 0 + a = a$ for all real numbers $a$. The student seems to be using the following:

$$a + b = a \qquad \text{where } a, b \text{ are real numbers.}$$

It would be interesting to find out if this student uses the same procedure on all addition problems.

19 The student is correct. The problem could be done as the student suggests.

$$\begin{array}{r l}
 & \overline{134 + 2 = 136} \\
25 & \big) 3420 \\
 & \underline{25\phantom{00}} \\
 & 92\phantom{0} \\
 & \underline{75\phantom{0}} \\
 & 170 \\
 & \underline{100} \\
 & 70 \\
 & \underline{50} \\
 & 20
\end{array}$$

When we show students the traditional algorithm, we usually gloss over this point.

20 One way of convincing the student that $-3 < -1$ would be as follows. Although it is true that a debt of three dollars is larger than a debt of one dollar, one might look at the situation in terms of how much money would be left over for oneself. Suppose that you had five dollars in your pocket. If you paid the person the three dollars that you owed that person, then you would have two dollars left over. If you paid the person the one dollar that you owed, then you would have four dollars left over for yourself.

Another way to convince a student that $-3 < -1$ would be to use the definition of less than. $-3 < -1$ because you must add a positive integer to $-3$ to get $-1$.

Another way to answer this question might be to think in terms of temperature. Three degrees below zero is colder than one degree below zero; i.e., there is less heat at three degrees below zero than at one degree below zero.

## 1.9 Number Theory

### Questions

1 While studying a unit on divisibility rules for numbers expressed in base-ten numerals, a student asks you if the divisibility rules for numbers expressed in other bases are the same. What is your reply?

2 You make the following statement to your seventh grade class. "A counting number $a$ is divisible by another counting number $b$ if $b$ divides $a$ evenly." What difficulty do you think students might have with this definition?

*How would you help seventh and eighth grade students to answer the following content questions?*

3 Isn't mod 7 arithmetic just like base 7?

4 "Why isn't 1 a prime number?"

5 An eighth grade student says that an even counting number is one which ends in 0, 2, 4, 6, or 8. Is this student correct?

6 A student rushes to your desk during class and tells you the following discovery: "If we take any two counting numbers, the greatest common factor multiplied by the least common multiple equals the product of the two numbers." Is this student correct?

7 A student in your eighth grade class asks, "What is the difference between the symbols $\div$ and $|$ when used in the following ways: $4 \div 8$ and $4|8$?" How would you respond?

8 The subject is divisibility rules. The divisibility rule for 8 is: if 8 can be divided into the number formed by the last three digits of a numeral, then 8 can be divided into the whole numeral evenly. The rule for 4 is the same, except we use the last two digits. Students know that if a number is divisible by 8 then it is also divisible by 4. A student tries the divisibility test for 8 and 8 doesn't work. He discovers that if the remainder is 4, then 4 is a divisor and he doesn't have to go through the test for 4. If the remainder is not 4, then 4 won't work either. Is the student right?

### Answers

1 For an answer to this question, read:

R.F. Wardrop, "Divisibility Rules for Numbers Expressed in Different Bases," *The Arithmetic Teacher*, March 1972, pp. 218-20.

2 This definition might lead to the following difficulty. Students might say that 6 is not divisible by 2 because $6 = 2 \cdot 3$ and 3 is not an even counting number. It would be better to leave out the word *evenly* in the definition.

3 Although there are certain similarities between mod 7 arithmetic and base 7, there are fundamental and essential differences. In the first place, base 7 is a numeration system. (Numeration systems are the actual systems used in writing numbers and numeration systems differ from one another in the way that numbers are named.) Mod 7 is a number system just as the set of integers, rational numbers, etc. are number systems. Number systems differ from one another in the behavior of the objects under discussion.

Mod 7 arithmetic consists of only a finite set of elements and no number is larger than 6; negative numbers and fractions are not needed. Base 7 consists of an infinite set of elements including negative and fractional numbers.

Thus, although there are certain similarities (such as using the numerals 0, 1, 2, 3, 4, 5, 6; the addition and multiplication identities, 0 and 1, are the same; many properties such as commutative, associative and distributive properties are valid in both systems), base 7 and mod 7 are formally different entitites.

4 At one time in history, the number "1" was considered to be a prime number. A major reason why the number "1" is no longer considered to be a prime number is because of the Fundamental Theorem of Arithmetic, which states that every composite positive integer can be factored uniquely into a product of primes. If "1" is allowed to be prime, then the uniqueness property is violated. Consider factoring 36 into a product of primes:

$$36 = 2^2 \cdot 3^2.$$

However, If "1" is considered a prime, then one can write

$$\begin{aligned} 36 &= 2^2 \cdot 3^2 \\ &= 2^2 \cdot 3^2 \cdot 1 \\ &= 2^2 \cdot 3^2 \cdot 1^2, \text{ etc.,} \end{aligned}$$

and the uniqueness is lost.

5 If the statement means that an even counting number when represented in base ten is a numeral whose last digit is 0, 2, 4, 6, or 8, then the student is correct. However, the numeration system used is important in the question. For example, in base five or in any odd base, $11_{\text{odd base}}$ represents an even counting number.

6 Yes, this student is correct. This question lends itself nicely to an inductive exploration such as given below.

Table 1.9.1 is put up on the blackboard and each student is asked to complete one line of the table. Then you might ask the students what they observe.

*TABLE 1.9.1*

| (1) | (2) | (3) | (4) | (5) | (6) |
|---|---|---|---|---|---|
| a | b | GCD (a, b) | LCM (a, b) | (3) X (4) | a X b |
| 1 | 2 | | | | |
| 2 | 3 | | | | |
| 3 | 6 | | | | |
| 4 | 12 | | | | |
| 5 | 9 | | | | |
| 8 | 16 | | | | |
| 8 | 24 | | | | |
| 16 | 84 | | | | |
| 20 | 90 | | | | |
| 30 | 120 | | | | |
| 36 | 81 | | | | |
| etc. | | | | | |

Most likely a formal proof that the GCD (a, b) multiplied by the LCM (a, b) equals the product of the two numbers, *a* and *b*, is beyond the grasp of your students, but an informal justification such as given below would not be.

Suppose we take a = 36 and b = 162 and factor each of them into prime representations.

$$36 = 2^2 \cdot 3^2$$
$$162 = 2 \cdot 3^4$$

$$36 \cdot 162 = (2^2 \cdot 3^2)(2 \cdot 3^4) = (2^2 \cdot 3^4)(2 \cdot 3^2)$$
$$= [\text{LCM}(36, 162)]\,[\text{GCD}(34, 162)]$$

One might then make the formal statement, using loose language, that the GCD takes care of those primes or powers of primes that are common to both numbers, while the LCM takes care of those primes or powers of primes that would be in the union of both numbers. Since in this process all the primes or powers of primes of the numbers are used up, then the result follows.

***Problem*** Construct a mathematical proof that LCM (a, b) · GCD (a, b) = a · b.

7 The symbol "÷" in 4 ÷ 8 usually means a binary operation and the result (in this case) is represented by the fraction $\frac{4}{8}$. The symbol "|" in 4|8 usually means a binary relation and stands for the statement that 4 divides 8 because 8 = 4 · 2.

8 Yes the student is correct. When the number represented by the last three digits is divided by 8, then the possibilities are that the remainders will be 0, 1, 2, 3, 4, 5, 6, 7. If the remainder is 0, then the number is divisible by 8. The only other time that the number is divisible by 4 is the situation in which the remainder is 4; i.e., the number represented by the last three digits is of the form 8k + 4 = 4(2k + 1). Since every other value after the hundreds place is divisible by 4, then the whole number must be divisible by 4.

# 1.10 Numeration Systems

## Questions

1 While studying different bases for numeration systems, one student asks if it is possible to have a negative number for a base. What do you tell this student?

2 An eighth grade student asks, "Is it possible to change $243_{seven}$ into a numeral in base three without first changing $243_{seven}$ into a numeral in base ten?" Can you help this student?

3 A student asks you if zero is a necessary element in our numeration system. What is your response?

4 In your seventh grade class, a student is trying to do the following problem.

$35_{twelve}$ multiplied by $46_{twelve}$ i.e.,

$$\begin{array}{r} 35_{twelve} \\ \times\ 46_{twelve} \\ \hline \end{array}$$

He knows that $5 \times 6 = 26_{twelve}$ but he can't figure out whether he should multiply $3 \times 6$ to get 16 and then add the 2, all in base twelve, or should he multiply $3 \times 6$ to get 18 and then add 2 to get 20 and then convert to base twelve. How would you handle this situation?

5 While studying a unit on different bases, a student asks you if it is necessary in, say, base 4 to use only the digits 0, 1, 2, 3, or if other digits can be used. What is your response?

6 You are teaching a unit on ancient systems of numeration, to include Roman numerals. An eighth grade student asks, "Since IV represents 4, why doesn't IXX represent 19 instead of XIX?" How do you explain the answer?

7 One student asks you how the Roman numerals represent fractional numbers. How would you answer this question?

8 For writing Roman numerals, the pupils are told the principles of economy, which was to use as few symbols as possible for each number. Therefore, 9 was written IX instead of VIIII. Johnny writes IIX instead of VIII for 8 and CCD instead of DCCC for 800. Do you tell him he is wrong or right, and why?

9 Your class asks you why 148 cannot be written as CIIL. How do you respond?

## Answers

1 For an answer to this question, read the following article by A.H. Nelson.* Also read:

Pedersen and Armbruster, "Models for Rational Number Base," *The Mathematics Teacher*, February 1975, p. 113.

# INVESTIGATION TO DISCOVERY WITH A NEGATIVE BASE

By **ALLYN H. NELSON**

Longwood High School
Middle Island, New York

CAN the elements of investigation and discovery that are so exhilarating to the research mathematician have the same effect on a high school student? I believe so. In fact, Professor G. Polya says, "A great discovery solves a great problem, but there is a grain of discovery in the solution of any problem. Your problem may be modest; but if it challenges your curiosity and brings into play your inventive faculties, and if you solve it by your own means, you may experience the tension and enjoy the triumph of discovery."[1] The challenge generally is to find a problem or area of investigation that might simulate in some small way the work of the research mathematician, and yet remain within the ability of the secondary student. I believe that an exploration into base negative two offers the student an opportunity to have some of these stimulating and illuminating experiences.

Why base negative two? The answer is very simple—more interesting things happen with base negative two than in other bases, interesting things that are within the grasp of junior and senior high school students. Aside from two very good articles in THE MATHEMATICS TEACHER[2] of February 1963, which delved into counting and addition in base negative two, I have found nothing to indicate any extensive work done in this area. I am especially indebted to Harry Foxwell, formerly a senior at Longwood High School and presently a student at Franklin and Marshall College, who was the first of my students to really dig into base negative two. He did such an excellent job and his results were so exciting that a paper he did on it won a first-place gold medal in the Long Island Mathematics Fair. Many of the examples and illustrations that follow are taken from the work he did and illustrate vividly the main point of this paper, i.e., base negative two is an excellent topic for student investigation.

The values of the positions in base negative two are shown in Figure 1. The first basic operation is counting, and in base negative two many very fascinating outcomes occur that can be observed by any youngster who has worked with negative numbers.

An example of the first few positive and negative numerals is shown in Table 1.

[1] *How to Solve It* (Garden City: Doubleday Anchor Books, Doubleday & Co., 1957), Preface to the First Printing, p. v.

[2] Richard D. Twaddle, "A Look at the Base Negative Ten," LVI, 88–90, and Chauncey H. Wells, Jr., "Using a Negative Base for Number Notation," LVI, 91–93.

*From *The Mathematics Teacher*, November 1967, pp. 723-26. Reprinted by permission.

| ... $(-2)^4$ | $(-2)^3$ | $(-2)^2$ | $(-2)^1$ | $(-2)^0$ | · | $(-2)^{-1}$ | $(-2)^{-2}$ | $(-2)^{-3}$ | $(-2)^{-4}$ ... |
|---|---|---|---|---|---|---|---|---|---|
| ↓ | ↓ | ↓ | ↓ | ↓ | | ↓ | ↓ | ↓ | ↓ |
| ... +16 | −8 | +4 | −2 | +1 | · | −1/2 | +1/4 | −1/8 | +1/16 ... |

FIGURE 1

The patterns that unfold in base negative two numeration can be discovered by most junior high students. In the units place, note the alternating order of one, zero, one, zero, etc., for both the positive and negative numbers. In the second position the pattern is one, one, zero, zero, one, one, etc. The pattern continues, doubling the number of repeating digits each time. Essentially a similar thing happens in base ten, of course, but the pattern is not nearly so evident. Another significant pattern is the number of digits in the positive numbers and the negative numbers. The positive numbers are represented by numerals with an odd number of digits and the negative numbers are represented by numerals with an even number of digits. A bit of exploration will quickly give the reason for this. The fact that the negative numbers are represented without the need of additional symbols raises some interesting possibilities and also some vexing problems. Explorations into "one-to-one correspondence" and "isomorphism" prove to be interesting for the more advanced student.

The operation of addition is relatively simple, but leads to some new methods.

TABLE 1

| POSITIVE | | NEGATIVE | |
|---|---|---|---|
| Base Ten | Base (−2) | Base Ten | Base (−2) |
| 1 | 1 | − 1 | 1 1 |
| 2 | 1 1 0 | − 2 | 1 0 |
| 3 | 1 1 1 | − 3 | 1 1 0 1 |
| 4 | 1 0 0 | − 4 | 1 1 0 0 |
| 5 | 1 0 1 | − 5 | 1 1 1 1 |
| 6 | 1 1 0 1 0 | − 6 | 1 1 1 0 |
| 7 | 1 1 0 1 1 | − 7 | 1 0 0 1 |
| 8 | 1 1 0 0 0 | − 8 | 1 0 0 0 |
| 9 | 1 1 0 0 1 | − 9 | 1 0 1 1 |
| 10 | 1 1 1 1 0 | −10 | 1 0 1 0 |
| 11 | 1 1 1 1 1 | −11 | 1 1 0 1 0 1 |
| 12 | 1 1 1 0 0 | −12 | 1 1 0 1 0 0 |
| 13 | 1 1 1 0 1 | −13 | 1 1 0 1 1 1 |
| 14 | 1 0 0 1 0 | −14 | 1 1 0 1 1 0 |
| 15 | 1 0 0 1 1 | −15 | 1 1 0 0 0 1 |
| 16 | 1 0 0 0 0 | −16 | 1 1 0 0 0 0 |
| 17 | 1 0 0 0 1 | −17 | 1 1 0 0 1 1 |
| 18 | 1 0 1 1 0 | −18 | 1 1 0 0 1 0 |
| 19 | 1 0 1 1 1 | −19 | 1 1 1 1 0 1 |
| 20 etc. | 1 0 1 0 0 | −20 etc. | 1 1 1 1 0 0 |

Since most problems in base negative two involve "carrying," the example in Table 2 should illustrate how the system works.

Observe that "carrying" involves the placing of the numeral one in the next two places, rather than in the next one as in positive bases. The reasons are simple to discover upon investigating the system. It is fascinating for a student to

TABLE 2

| Base Ten | Base (−2) 1st Carry | Base (−2) 2nd Carry | Base (−2) 3rd Carry | Base (−2) Note* |
|---|---|---|---|---|
| | | | 1 1 | 1 1 |
| | | 1 1 | 1 1 | 1 1 |
| | 1 1 | 1 1 | 1 1 | 1 1 |
| 1→ | 1→ | 1→ | 1→ | ↓ 1 |
| +3→ | +1 1 1→ | + 1 1 1→ | + 1 1 1→ | ↓ 1 1 1 |
| 4→ | 0→ | 0 0→ | 1 0 0→ | 0 1 0 0 |

* Note that the sum of either of the combinations at the right will equal zero. A look into the structure of the system will verify it. This pattern will come in handy later.

| 1 1 | 1 |
|---|---|
| + 1 | + 1 1 |
| 0 | 0 |

note here that a number plus its additive inverse will equal zero, although the numerals would seem to indicate differently. It is significant to note that the addition of four ones in one column are equivalent to a single one in the column two places to the left. A look into the structure of the system will quickly verify this.

Subtraction is also a most fascinating operation in base negative two, simply because there are a number of ways of performing it. Examples are found in Tables 3 and 4.

TABLE 3

| Base Ten | "Exchange" Base (−2) | | "Additive Inverse" Base (−2) | |
|---|---|---|---|---|
| 2 | 1 1 0→ | 1̸ 1̸ (110) | 1 1 0→ | 1 1 0 |
| − 1 | − 1→ | − 1 | − 1→ | + 1 1 |
| 1 | | 1 | | 0 1 |

A little study of the examples above will give the reader an insight into the method used in each. It might be well to observe here that the common exchange method ("borrowing") will work only if there is something to "borrow" from. (The reader might try $4_{\text{ten}} - 2_{\text{ten}}$ in base negative two to verify it.) The "replacement" method above appears to be unique to a negative base. It would seem that the "additive inverse" and the "equal additions" methods are sounder mathematics than the others.

Multiplication entails only repetitive addition. Once addition has been mastered, multiplication is relatively easy, as indicated in the example below:

```
   5 →        1 0 1
 × 2 →      × 1 1 0
 -----      -------
            1 0 1 0
          1 0 1
          ---------
  10 →    1 1 1 1 0
```

The younger student might well find the product of two negative numbers and the product of a positive and a negative of some interest.

A stimulating area of discovery opens up when we attempt division. It will be found that our more sophisticated division algorithms are extremely difficult when working in base negative two, whereas the so-called "modern" method is relatively easy. Using the "modern" method to divide twelve by two, we get:

```
1 1 0 / 1 1 1 0 0
        1 1 0 0 0  |    1 0 0 → (+4)
        ---------  |
            1 0 0  |
          1 1 0 0  |      1 0 → (−2)
        ---------  |
        1 1 0 0 0  |
        1 1 0 0 0  |    1 0 0 → (+4)
        ---------  |  ---------   ----
                0  | 1 1 0 1 0 → (+6)
```

If we attempt the problem using our more common division algorithm, we run into a number of difficulties. Using the problem mentioned above for an example, we get:

```
              1 1
1 1 0 / 1 1 1 0 0
        1 1 0
        -----
          1 0
          1 1 0
          -----
          1 0 0
```

At this point we must stop, since the re-

TABLE 4

| "Replacement" Base (−2) | | "Equal Additions" Base (−2) | | "Complement" Base (−2) | |
|---|---|---|---|---|---|
| 1 1 0 → | 1 1 0 (boxed 1 1 above, 1) | 1 1 0 → | 1 1 (110 above) | 1 1 0 → | 1 1 0 |
| − 1 → | − 0 | − 1 → | − 1 (1 1 above) | − 1 → | +1 0 0 0 1 |
| | 0 1 | | 0 0 1 | | 1̸ 0 0 0 1 |

mainder (1 0 0) is larger than the divisor.

```
         1 0 1
1 1 0 / 1 1 1 0 0
        1 1 0
        -----
            1 0 0
            1 1 0
            -----
            1 1 0
```

Again a problem, since the remainder is equal to the divisor. Starting at the third position leads nowhere; therefore, we are left with the alternative of trying another. If we start in the second position, we get:

```
          1
1 1 0 / 1 1 1 0 0
        1 1 0
        -----
      1 1 0 1
```

This time the remainder is a negative number whereas the divisor is positive.

```
        1 1 0 1 0
1 1 0 / 1 1 1 0 0
        1 1 0
        -----
          1 1 1
          1 1 0
          -----
            1 1 0
            1 1 0
            -----
                0 0
```

Even in this step there were some complications that might have been difficult. Rather than continue for a long time, I terminated the problem, since the main idea has been demonstrated sufficiently. This method has been done in a purely mechanical way, but if one stops to think about it, is not this the way most of us do division? A great deal of insight into the system would, of course, enable us to use the more sophisticated **algorithm** above.

Some amazing results are obtained when attempting division of rational numbers. One startling result is the nonuniqueness of expressions of certain rational numbers. Uniqueness is a property of so many mathematical structures that we have a tendency to overlook its importance. In the case of base negative two, the nonuniqueness is, of course, in the representation of certain numbers, and not in the numbers themselves. In base negative two, the rational number 1/111 ($1/3_{\text{ten}}$) can be represented as $0.010\overline{101}$ or as $1.101\overline{010}$. A quick check, using the formula for the sum of an infinite geometric sequence, will show that both equal 1/111. The student investigating base negative two might well note here the distinction between a property of number such as uniqueness and the nonuniqueness noted, which is a property of the system of numeration. It is mystifying to see that 1/111 ($1/3_{\text{ten}}$) is the only "proper" rational number with two ways of expressing it. The proof of this is relatively easy if one observes that the maximum value of any "bicimal" expansion is $.010\overline{101}$ or 1/111. The only other way of expressing a "proper" rational is to use "1" along with a negative "bicimal" expansion. The maximum negative bicimal expansion is $.101\overline{010}$ or 110/110 ($-2/3_{\text{ten}}$), which when combined with "1" is $1.101\overline{010}$ or 1/111 ($1/3_{\text{ten}}$). The so-called "improper" rationals 100/111 ($4/3_{\text{ten}}$), 11011/111 ($7/3_{\text{ten}}$), 11110/111 ($10/3_{\text{ten}}$), etc., can also be expressed in two ways (100/111 is $110.101\overline{010}$ or $1.010\overline{101}$). The proof is similar to that above.

It is also possible to express base negative two using scientific notation. For example:

$$110000000 = 1.1 \times 10^{11000}, \qquad [11000 = 8_{\text{ten}}]$$
$$0.000000011 = 1.1 \times 10^{1000}, \qquad [1000 = -8_{\text{ten}}]$$

An attempt to construct logarithms with negative two as a base leads to some rather interesting events. We run into discontinuities, nonexistence, and the rather startling result of a real value for the logarithm of a pure imaginary. E.g.,

$$\text{Log}_{(-2)}\, 2i\sqrt{2} = 3/2 \qquad \text{since} \qquad (-2)^{3/2} = 2i\sqrt{2}.$$

An example of nonexistence is

$$\text{Log}_{(-2)}\, 8 = ?$$

I am sure that investigations by others

will lead to many more discoveries, maybe something of value about base negative two, but if these considerations do nothing more than challenge curiosity, this paper will have been of value. Investigation into prime numbers and repeating decimals with base negative two will lead students to distinguish between a property of number (prime) and a property of the system of numeration, such as terminating and nonterminating decimals (or "bicimals"). Many explorations will yield nothing of value, but the student will gain a more mature insight into the structure of mathematics and—just as important—have some fun in doing so.

2 There are various algorithms that can be employed to do the problem. One algorithm is given below and other algorithms, plus the one presented here, can be found in the references listed.

The algorithm presented is a generalization of a common algorithm that is used to change a numeral in base ten into a numeral in some other base.

Change $125_{ten}$ into a numeral in base seven.

$$\begin{array}{rl} 7\,\underline{|\,125} & \quad 6 \\ 7\,\underline{|\,17} & \quad 3 \\ 7\,\underline{|\,2} & \quad 2 \\ 0 & \end{array} \qquad (1)$$

Thus $125_{ten} = 236_{seven}$.

Change $243_{seven}$ into a numeral in base three.

$$\begin{array}{rl} 3\,\underline{|\,243_{seven}} & \quad 0 \\ 3\,\underline{|\,61_{seven}} & \quad 1 \\ 3\,\underline{|\,20_{seven}} & \quad 2 \\ 3\,\underline{|\,4_{seven}} & \quad 1 \\ 3\,\underline{|\,1_{seven}} & \quad 1 \end{array} \qquad (2)$$

Thus $243_{seven} = 11210_{three}$.

The procedure in (2) is the same as the procedure in (1) except that all computations in (2) are in base seven.

It might be instructive to look at the following example, which is a little more complicated.

Change $11210_{three}$ into a numeral in base seven.

Since all computations are to be done in base three, then the first step is to rewrite 7 in base three as 21.

$$\begin{array}{rl} 21_{three}\,\underline{|\,11210_{three}} & \quad 10_{three} = 3_{seven} \\ 21\,\underline{|\,200_{three}} & \quad 11_{three} = 4_{seven} \\ 21\,\underline{|\,2_{three}} & \quad 2 \quad = 2_{seven} \\ 0 & \end{array}$$

Thus $11210_{three} = 243_{seven}$.

Thus, if one is changing a numeral in base x into a numeral in base y, then the procedure is to divide y (expressed in base x) into the numeral in base x where all computations are in base x.

Additional algorithms can be found in the references listed below.

(1) M.J. Barrett, "A method for changing numerals in certain nondecimal bases to numerals in other certain nondecimal bases, directly." *The Arithmetic Teacher*, May 1968, pp. 453–54.

(2) James Husk, "Changing bases without using base ten." *The Arithmetic Teacher*, May 1968, pp. 461-62.

(3) John Lee, "Changing bases by direct computation." *The Mathematics Teacher*, December 1972, pp. 752-53.

(4) Joe Smith, "A method for converting from one nondecimal base to another." *The Arithmetic Teacher*, April 1968, pp. 344-46.

3 For an answer to this question, read the following article by Charles Brumfiel.* Also read:

Boyd Henry, "Zero, the Troublemaker," *The Arithmetic Teacher*, May 1969, p. 365.

# Zero is highly overrated

CHARLES BRUMFIEL

*University of Michigan, Ann Arbor, Michigan*

*Mr. Brumfiel is a professor in the mathematics department at the University of Michigan.
He is well known to teachers of mathematics through the Ball State Experimental Project.*

It is instructive to consider a place value system without a zero. We modify our base ten system by writing *t* for ten and agreeing to count

1, 2, 3, 4, 5, 6, 7, 8, 9, *t*, 11, 12, . . . , 19, 1*t*, 21, . . . , 9*t*, *t*1, . . . , *tt*, 111, 112, . . . .

Notice that one hundred is written as 9*t* (nine tens and ten ones). The symbol *tt* represents ten tens and ten ones, or one hundred ten. If a "*t*" does not occur in a number symbol the meaning of the symbol is the familiar one: 342 is still

3 hundreds + 4 tens + 2 ones.

In expanded notation, and using *t* for ten, we have

$$342 = 3 \cdot t^2 + 4 \cdot t + 2.$$

Any "old-fashioned" number symbol can be easily replaced by a symbol in this new notation. The reasoning is no more difficult than that used when we "borrow" in subtraction. We are accustomed to think of 200 as

100 + 90 + 10

and we have

| *Old Symbol* | | *New Symbol* |
|---|---|---|
| 200 | = | 19*t*. |

The reader should verify several relationships like the following:

*From *The Arithmetic Teacher*, May 1967, pp. 377-78. Reprinted by permission.

| *Symbol in Base Ten with Zero* | | *Symbol in Base Ten without Zero* |
|---|---|---|
| 40 | = | $3t$. |
| 5,000 | = | $499t$. |
| 70,302 | = | $6t2t2$. |

Let's take a look at computation in this numeration system. First we should make sure that we really can count—

| | |
|---|---|
| After 99 | comes $9t$. |
| After $3tt$ | comes 411. |
| Before 371 | comes $36t$. |
| After $tttt$ | comes 11,111. |
| Before 74,111 | comes $73ttt$. |

Addition and subtraction techniques in this numeration system require very little modification of our usual habits. We "carry" and "borrow" tens, hundreds, etc., just as before. However, we can now record ten units in any place and we *must* record at least one in each place.

$$\begin{array}{r} 43 \\ +\ 27 \\ \hline 6t \end{array} \qquad \begin{array}{r} 87 \\ +\ 23 \\ \hline tt \end{array} \qquad \begin{array}{r} t65 \\ +\ 844 \\ \hline 18t9 \end{array} \qquad \begin{array}{r} 1ttt \\ +\ \ ttt \\ \hline 321t \end{array}$$

$$\begin{array}{r} 65 \\ -\ 25 \\ \hline 3t \end{array} \qquad \begin{array}{r} 511 \\ -\ 24t \\ \hline 261 \end{array} \qquad \begin{array}{r} 432. \\ -\ \ tt \\ \hline 322 \end{array}$$

Notice that in the first subtraction problem we had to borrow one of the 6 tens since we can't record the difference $5 - 5$.

A subtraction problem like

$$\begin{array}{r} 3111 \\ -\ \ ttt \\ \hline \end{array}$$

deserves a little special thought. The reader should enjoy testing his understanding by seeing how quickly he can produce the answer.

As preparation for multiplication we should first practice counting by tens.

$$t,\quad 1t,\quad 2t,\quad \ldots,\quad 14t,\ \ldots$$
$$1 \cdot t,\ 2 \cdot t,\ 3 \cdot t, \ldots, 15 \cdot t, \ldots$$

Notice that $15 \cdot t$ or fifteen tens is fourteen tens and ten ones,

$$15 \cdot t = 14t.$$

The rule for multiplying by ten should be clear: 7652 tens is 7651 tens + 1 ten.

$$7652 \cdot t = 7651t.$$

Multiplying by any number less than $t$ is straightforward.

$$\begin{array}{r} 25 \\ 2 \\ \hline 4t \end{array} \qquad \begin{array}{r} 36 \\ 5 \\ \hline 17t \end{array} \qquad \begin{array}{r} 42t \\ 7 \\ \hline 29tt \end{array}$$

Now that we can multiply by $t$ and by any number less than $t$ we can combine these skills and work more complicated multiplication problems. Since $12 = t + 2$ we can work the multiplication problem below as shown.

$$\begin{array}{rl} 32t & \\ 12 & \\ \hline {}^{*}65t & = 2 \cdot 32t. \\ 329t & = t \cdot 32t. \\ \hline 395t & \end{array}$$

For a product like $35 \cdot 346$ we have

$$35 \cdot 346 = (3 \cdot t + 5) \cdot 346.$$
$$= 3 \cdot t \cdot 346 + 5 \cdot 346.$$

The work is shown below. On the side we calculate $3 \cdot t \cdot 346$.

$$\begin{array}{rl} 346 & \\ 35 & \\ \hline 172t & = 5 \cdot 346 \\ t37t & = 3 \cdot t \cdot 346 \\ \hline 11ttt & = (3t + 5) \cdot 346. \end{array} \qquad \begin{array}{rl} 345t & = t \cdot 346 \\ 3 & \\ \hline t37t & = 3 \cdot t \cdot 346. \end{array}$$

For larger multipliers we must multiply repeatedly by $t$.

$$\begin{array}{rl} 4t2 & \\ 326 & \\ \hline 2t12 & = 6 \cdot 4t2 \\ 9t3t & = 2 \cdot t \cdot 4t2 \\ 14t59t & = 3 \cdot t^2 \cdot 4t2 \\ \hline 163{,}652 & = (3t^2 + 2t + 6) \cdot 4t2. \end{array}$$

The second and third partial products are computed thus:

$$\begin{array}{rl} 4t1t & = t \cdot 4t2 \\ 2 & \\ \hline 9t3t & = 2 \cdot t \cdot 4t2 \end{array} \qquad \begin{array}{rl} 4t1t & = t \cdot 4t2 \\ 4t19t & = t^2 \cdot 4t2 \\ 3 & \\ \hline 14t59t & = 3t^2 \cdot 4t2. \end{array}$$

Multiplication by ten and by powers of ten is so simple we could do the later calculations mentally. We wouldn't think of permitting our students to use such a crutch in multiplying (would we?).

The reader is urged to practice multiplication until he can abandon the crutch.

* There are excellent opportunities to "cheat" here. You can think: "$32t$ is really 330 and $2 \cdot 330$ is 660. I'll write this as $65t$."

Check the examples below.

```
  34              513
  25              286
 ---             ----
 16t             2t78
 67t            3tt3t
 ---           t259t
 84t          -------
              146,718
```

A division example is worked below by a repeated subtraction process.

```
32)45t8
   319t = t² · 32
   ----
   13t8
   127t = 4 · t · 32
   ----
    128
    128 = 4 · 32.
   ----
```

The quotient is $1 \cdot t^2 + 4 \cdot t + 4 = 144$.

The reader might find the following an interesting exercise. Construct a base-three system without zero. It is natural to use the symbols 1, 2, 3, and count

1, 2, 3, 11, 12, 13, 21, 22, 23, 31, 32, 33, 111, 112, . . .

but any three distinguishable symbols will do as well. Formulate and work some problems in this notation.

In conclusion, we remark that we had always considered the symbol for zero an essential feature of any place value system. Clearly, this is not the case. Zero is indeed highly overrated. It isn't even needed as a placeholder!

4 It really does not make any difference since the answer would be the same in any case. If the student multiplies $3 \times 6$ to get 18, adds 2 to get 20 (all in base ten) and then converts to base twelve, the student would get 18 in base twelve. This is the same answer the student would get working strictly in base twelve. In general, whether one works directly in the given base or works in base ten and then converts, the same answer would be obtained in either case.

5 For an answer to this question, read the following article by Norman Woo.* Also read:

J. Jordan, "Small-Digit Representation of Real Numbers," *The Mathematics Teacher*, January 1968, p. 36.

# A GENERALIZED BASE FOR INTEGERS

By **NORMAN WOO**

**California State University at Fresno**
**Fresno, California**

ONE of the most important notations in mathematics, the positional system of notation of numbers, seems to have been invented in India by the Hindus. It was certainly transmitted to the Western world by the Arabs in about the eighth century A.D. At first the Hindu-Arabic system of notation was of interest only in the marketplace, but eventually it came to be of interest to mathematicians, and it is still the object of considerable study.

It is well known that the set of nonnegative integers can be represented by selecting an integer, $n > 1$, as a base. The digits 0, 1, 2, 3, . . . , $n - 1$ are then used, and with these any nonnegative integer $m$ can be uniquely represented as

$$m = a_t n^t + a_{t-1} n^{t-1} + \cdots + a_1 n^1 + a_0$$

where are $a_i$'s are digits.

*From *The Mathematics Teacher*, February 1973, pp. 169-70. Reprinted by permission.

Some effort was made a few years ago to study representations of integers when the base was $n < -1$. In particular, negative two was suggested by N. G. De Bruijn (1950) and by Chauncey H. Wells, Jr. (1963), and then negative ten by Richard D. Twaddle (1963). The digits used in the representation for base negative two were 0 and 1; 0, 1, 2, 3, 4, 5, 6, 7, 8, 9 were used in the representation for base negative ten. James Jordan (1968) generalized these concepts of representation to an arbitrary base $h$ (a positive or negative integer) and chose as the digits the integers $x$ where

$$-\frac{|h|}{2} < x \leq \frac{|h|}{2}.$$

However, he did not give the complete details of the representation for any arbitrary integer $n$. The purpose of this paper is to give some examples to show how one deals with the actual representation.

For example, if we take $h = 16$ as a base, then the digits for representation would be taken from the set

$$A = \{-7, -6, -5, -4, -3, -2, -1, 0, 1, 2, 3, 4, 5, 6, 7, 8\}.$$

In order to avoid any confusion between the subtraction of an integer $n$ and the negative integer $n$ used in the representation, we adopt the following notation for the set

$$A = \{\bar{7}, \bar{6}, \bar{5}, \bar{4}, \bar{3}, \bar{2}, \bar{1}, 0, 1, 2, 3, 4, 5, 6, 7, 8\}.$$

If we want to express a number 1769 in base 16, we apply the division algorithm:

$$1769 = 110(16) + 9.$$

But 9 is not an element of set $A$. However, 9 can be expressed in the following form:

$$9 = 1(16) + (\bar{7}).$$

Therefore,

$$\begin{aligned} 1769 &= 110(16) + 1(16) + (\bar{7}) \\ &= 111(16) + (\bar{7}). \end{aligned}$$

Furthermore,

$$111 = 7(16) + (\bar{1}).$$

Finally, we have

$$\begin{aligned} 1769 &= [7(16) + (\bar{1})](16) + (\bar{7}) \\ &= 7(16)^2 + \bar{1}(16) + (\bar{7}). \end{aligned}$$

Or,

$$1769 = 7\bar{1}\bar{7} \text{ in base } 16.$$

Let us now express an integer in a negative base. For example, if $h = -7$, the corresponding set of digits would be

$$B = \{\bar{3}, \bar{2}, \bar{1}, 0, 1, 2, 3\}.$$

If we want to express the number 173 in this base, we have

$$\begin{aligned} 173 &= (-25)(-7) + (\bar{2}) \\ -25 &= (4)(-7) + (3) \\ 4 &= (-1)(-7) + (\bar{3}). \end{aligned}$$

Finally,

$$\begin{aligned} 173 &= ([(-1)(-7) + (-3)](-7) + (3))(-7) + (\bar{2}) \\ &= ((\bar{1})(-7)^2 + (\bar{3})(-7) + (3))(-7) + (2) \\ &= (\bar{1})(-7)^3 + (\bar{3})(-7)^2 + (3)(-7) + (\bar{2}) \\ &= \bar{1}\bar{3}3\bar{2} \text{ in base } -7. \end{aligned}$$

Note that the base and its digits are very much interrelated with each other. In fact, any integer $h$, other than 1, 0, −1, can be used as a base if we select its digits $x$ such that

$$-\frac{|h|}{2} < x \leq \frac{|h|}{2}.$$

The purpose of this paper is not to encourage students to adopt "nonten" bases, but rather to promote the understanding of the base ten system. It is hoped that work with bases other than ten will increase understanding as well as improve skills for both secondary school mathematics teachers and students.

#### REFERENCES

De Bruijn, N. G. "On Bases for the Set of Integers." *Publication Mathematics Debrecen* 1 (1950): 232–42.

Jordan, James H. "Small-digit Representation of Real Numbers." MATHEMATICS TEACHER 61 (January 1968): 36–38.

Twaddle, Richard D. "A Look at Base Negative Ten." MATHEMATICS TEACHER 56 (February 1963): 88–90.

Wells, Chauncey H., Jr. "Using a Negative Base for Number Notation." MATHEMATICS TEACHER 56 (February 1963): 91–93.

6 The Roman numeration system essentially is an additive system with subtractive and multiplicative features. If symbols decrease in value from left to right, then their values are added. If a symbol has a smaller value than a symbol on the right, then the first symbol is subtracted from the second symbol.

Also, the Romans invented their system with the restriction that no more than two symbols were involved in the subtractive feature and not more than three identical symbols were used in succession. The first feature was used probably to avoid ambiguity in expression. For example, what would IVX represent? In one case, IVX could mean I(VX) = I(V) = 5 – 1 = 4. In another case, IVX could mean (IV)X = 10 – 4 = 6. Since the Romans did not use parenthesis, which answer would be the correct one? The second feature was probably used for economy in order to use fewer symbols.

Thus, in the student's question, using the subtractive features, IXX would represent X – (IX) = 10 – 9 = 1, not 19.

7 Roman fractions were used in base twelve. Special symbols and words were used for

$$\frac{1}{12}, \frac{2}{12}, \ldots, \frac{11}{12}, \frac{1}{24}, \frac{1}{36}, \frac{1}{48}, \ldots$$

Base twelve was used probably because of the relation of the lunar month to the year. The Roman unit of weight was the "As" which weighed one pound and was divided into 12 uniciae or "ounces." Successive parts were divided into halves and thus fractional units of 12ths, 24ths, etc. were involved. The Romans also used an abacus for computation. The abacus had 2 grooves for fractions. One groove measured units of twelve and the second groove represented fractions of 24ths, 36ths, etc.

8 Johnny is correct in trying to use as few symbols as possible, and he is doing some good thinking. However, he is violating a condition of the Roman numeration system, which states that no more than two symbols are to be involved in the subtractive feature of this numeration system. (See discussion in Answer 6 of this section for further information.)

9 If 148 was written as CIIL, then this notation would violate the condition that no more than two symbols be involved in the subtractive feature of the Roman numeration system. (See discussion in Answer 6 for further information.)

## 1.11 Percent

### Questions

1 You have just given your class the following problem to solve. "Find 28% of 50." You call on one of your students to put his solution on the blackboard. Below is the student's solution.

```
  28
× .5      28% of 50 is 14.
  14
```

Is this student correct?

2 A student in your class asks, "Why do banks, finance companies, loan sharks and credit unions all charge different interest rates on loans?" How do you respond?

3 If a student in your class asks, "Why is there no 100th percentile on standardized tests?" how would you answer?

4 "My father bought a house for $20,000 at 12% interest. He told me that by the time he finishes paying for the house, it will have cost him more than $50,000. How is that possible? 12% of 20,000 is only $2,400." If a student presented you with this question, how would you respond?

5 You ask the class to find 62% of 50. A student turns in the following solution.

```
  9
 1ø0%        50       50
 -62%       ×.38     -19
  38%       400      31%
           150
           19.00
```

How do you help this student?

6 You ask your seventh grade class to find $\frac{1}{10}$% of 50. One student writes that $\frac{1}{10} = .1$, then $\frac{1}{10}$% of 50 equals $(.1)(50) = 5$. How would you help this student.

### Answers

1 On the surface it looks like the student has the correct answer but the wrong multiplication. It seems like the student was solving the problem of finding 50% of 28. You would probably want to question him to find out what he was doing. In any case, an unexpected situation has occurred, which an alert teacher might want to explore. In this particular case, it seems as if 28% of 50 is equal to 50% of 28. The

question that immediately comes to mind is the following: "Is x% of y equal to y% of x?" A teaching strategy that one might use in this case is to ask each student to make up five different problems on his own and test the above hypothesis. After this exploratory work is completed, you might ask if anyone found a case that does not work. Assuming that each student has multiplied correctly, the answer would be no. Thus it seems as if the above statement is true, and you might ask if anyone can devise a proof of the above hypothesis. Not only do students get meaningful practice using this strategy, but they also get involved in a neat discovery situation.

A proof of the statement that x% of y is equal to y% of x might run along the following lines:

$$x\% \text{ of } y = \frac{x}{100} \cdot y = \frac{x \cdot y}{100} = \frac{y \cdot x}{100} = \frac{y}{100} \cdot x = y\% \text{ of } x.$$

2 In effect one is paying a certain amount of money for the privilege of borrowing money; i.e., one is really *buying* money. Just as the same brand of a product such as a television set has different prices in different stores, so the buying of money from different companies has different prices. There are different procedures that various finance agencies use to figure out how much they will charge for borrowing money such as the time payment plan, revolving charges, unpaid balances, etc. It might be interesting to have the students find out about these various methods and discuss them in class.

3 The *P*th percentile of a group of scores is defined to be the point below which *P* percent of the scores are. Thus if the 100th percentile of a group of scores was calculated, what would this point be and what information would it give? For example, suppose that the following represent scores received by a group of students on a given test:

25 40 70 85 90 92.

What would be the 100th percentile of this group of scores: 93, 92.5, 94, 95, etc.? In fact, any real number greater than 92 would be the 100th percentile. What information does this give someone? All that this really says is that all of the scores are below a certain point, but one could readily get this information by looking at the highest score. Thus no 100th percentile is computed.

4 The student is, of course, using simple interest for one year and must realize that the mortgage stretches over a longer period of time, usually 20 or 30 years. In this case only part of the debt is paid off (a small part) during the first year; thus interest is computed on the amount of money still owed. The usual home mortgage requires that a person make equal monthly payments, part of which is used for paying the interest on the outstanding debt and part of which is applied to reducing the outstanding debt. For figuring out the mortgage payments mathematically, one has to use a geometric series. Whether a teacher should go into a detailed explanation of how a mortgage is actually determined at this level is doubtful. However, if one wanted to delve into such an explanation, the following paper shows how this is done (C. Sloyer, "Fantastiks of Mathematiks").

*THAT #%&*? MORTGAGE PAYMENT*
*Mathematics Used: Algebra, Geometric Series*

Suppose that an individual obtains a $30,000 mortgage at an interest rate of 6% for a period of 20 years. (This interest rate is not very realistic in 1976.) Each month, the individual is to make a payment of P dollars (the same each month). We show here how to determine the value of the number P.

At the end of the first month, the individual pays P dollars (the first payment). A part of this, say $P_1$, is on the principal, while $\frac{.06}{12}(30{,}000) = .005(30{,}000) = 150$ is for interest. We thus see that $P > \$150$. We write

$$P = \underbrace{P_1}_{\text{on principal}} + \underbrace{(30{,}000).005.}_{\text{on interest}} \tag{1}$$

In order to simplify what follows, let i = .005, so that (1) becomes

$$P = P_1 + (30{,}000)i. \tag{2}$$

At the end of the second month, the payment P is $P_2$ dollars on the principal plus interest on the remaining principal $30{,}000 - P_1$. Thus,

$$P = P_2 + (30{,}000 - P_1)i. \tag{3}$$

Continuing, at the end of the third month, we have

$$P = P_3 + (30{,}000 - P_1 - P_2)i, \tag{4}$$

and at the end of the fourth month

$$P = P_4 + (30{,}000 - P_1 - P_2 - P_3)i. \tag{5}$$

In general, at the end of month k, we have

$$P = P_k + (30{,}000 - P_1 - P_2 - \cdots - P_{k-1})i. \tag{6}$$

Now, from (2) and (3), we have

$$P_1 + (30{,}000)i = P_2 + (30{,}000 - P_1)i$$

or

$$P_2 = P_1(1 + i). \tag{7}$$

From (3) and (4), we have

$$P_2 + (30{,}000 - P_1)i = P_3 + (30{,}000) - P_1 - P_2)i$$

or

$$P_3 = P_2(1 + i),$$

which, together with (7), yields

$$P_3 = P_1\ (1 + i)^2. \tag{8}$$

From (4) and (5), we have

$$P_4 = P_3(1 + i),$$

which, together with (8), yields

$$P_4 = P_1 (1 + i)^3. \tag{9}$$

Summarizing, we see

$$\left.\begin{aligned} &P_1 \\ &P_2 = P_1 (1 + i) \\ &P_3 = P_1 (1 + i)^2 \\ &P_4 = P_1 (1 + i)^3 \end{aligned}\right\} = P \tag{10}$$
$$\vdots$$

This pattern continues (the student can show this by induction) so that

$$P_k = P_1 (1 \pm i)^{k-1}.$$

Now, $P_1$, $P_2$, $P_3$, etc. are payments on the principal and so

$$P_1 + P_2 + P_3 + \cdots + P_{240} = 30{,}000$$

or, using the summary in (10) and the general term, we have

$$P_1 + P_1 (1 + i) + P_1 (1 + i)^2 + \cdots + P_1 (1 + i)^{239} = 30{,}000.$$

The left hand side of the above equation is a geometric series. Using the formula for the sum of such series, we have

$$\frac{P_1 [(1 + i)^{240} - 1]}{(1 + i) - 1} = 30{,}000 \quad \text{or} \quad P_1 = \frac{(30{,}000)i}{(1 + i)^{240} - 1}.$$

Using interest tables, logarithms, or a calculator, one obtains

$$P_1 = 64.93 \text{ (to the nearest cent).}$$

So that (from (1)) we have

$$P = 64.93 + 150 \quad \text{or} \quad P = \$214.93.$$

***Problem*** Determine the monthly payments on a $40,000 mortgage at an interest rate of 9% for a period of 25 years.

5 The method that this student is using is a round about but correct way of doing this problem. The student should be commended for devising this unusual method, assuming, of course, that she understands what has been done. You would also want to know why she used this method and then show her the standard way of doing the problem, which involves less work and less chance of error.

6 This is a common error; one way of helping the student is:

$$\frac{1}{10}\% = \frac{1}{10} \text{ of one percent, i.e., } \frac{1}{10} \cdot (1\%).$$

Since $1\% = \frac{1}{100}$, then

$$\frac{1}{10} \cdot (1\%) = \frac{1}{10} \cdot \frac{1}{100} = \frac{1}{1000}.$$

Thus $\frac{1}{10}$% of 50 is

$$\frac{1}{1000} \cdot 50 = \frac{50}{1000} = .05.$$

## 1.12 Sets

### Questions

1 A seventh grade student says that he has another definition of subset: "If set A does not contain an element that is not also an element of set B, then A is a subset of B." Is the student correct?

2 A seventh grader says that $\{\emptyset\}$ is the notation that you should use for the empty set. What is your response?

3 "If $A = \{1, 2, 4\}$ and $B = \{1, 2, 3\}$, why isn't $A \cup B = \{1, 1, 2, 2, 3, 4\}$? Or, if A = {science book, math book} and B = {French book, math book}, isn't $A \cup B$ = {math book, math book, science book, French book}?" This question frequently arises in the classroom. What is your reply?

4 You are going over the following problem in class. "Indicate whether the following statement is true or false:

If $A \cap B = A \cup B$, then $A \subseteq B$."

One student says that it is false because A would have to be equal to B, not a subset of B. Is the student right? How would you answer?

5 Your class is working on the following problem. "Analyze the data below and find out how many students are in the school. Assume that every student in the school takes one of the courses."

28 students take English
23 students take French
23 students take German
12 students take English and French
11 students take English and German
8 students take French and German
5 students take all three courses

One of your students does the problem as follows.

28
23
23
74 students total if students took only one course,

but 12
8
11
31 taking 2 courses
−5 taking 3 courses
26 taking more than one course. 74 − 26 = 48 students in the school.

Is this student correct in the method used and answer given?

## Answers

1 The student is correct. Either $A \subseteq B$ or $A \nsubseteq B$. If $A \nsubseteq B$, then A must contain at least one element that is not in B. The student is saying that this is not the case. Therefore, $A \nsubseteq B$ cannot happen, and thus A must be a subset of B.

2 The student is incorrect. Probably the best way to convince the student is to make the analogy that the empty set is similar to an empty paper bag. Thus $\{\emptyset\}$ would be classified as an empty paper bag inside another empty paper bag. Thus the second paper bag contains something inside it – another empty paper bag.

3 For an answer to this question, read:

(1) I. Feinstein, "Sets: A Pandora's Box?", *School Science and Mathematics*, June 1973, pp. 495-500.

(2) Geddes and Lipsey, "The Hazards of Sets," *The Mathematics Teacher*, October 1969, p. 454.

4 The student is incorrect, but he does have the correct idea, that A would have to be equal to B. However, the main difficulty that the student has here is with the notation $A \subseteq B$. This notation means A is either a proper subset of B *or* A is equal to B. Since the last alternative is certainly the case in this situation, then the notation $A \subseteq B$ may be used.

5 The student is correct in both the method used and the answer given. The student is intuitively using the well-known formula:

$$n(A \cup B \cup C) = n(A) + n(B) + n(C) - n(A \cap B) - n(A \cap C) - n(B \cap C) + n(A \cap B \cap C)$$

where n(S) is the number of elements in set S. Thus,

$$\begin{aligned} n(A \cup B \cup C) &= (28 + 23 + 23) - (12 + 8 + 11) + 5 \\ &= 74 - 26 \\ &= 48. \end{aligned}$$

## 1.13 Square Root

### Questions

1 After studying the extraction of square roots, a student asks, "Can we extract cube roots?" How would you answer this student?

2 A student asks, "If it is true that the square root of 25 is ±5, then why doesn't $\sqrt{25} = \pm 5$?" How do you reply?

### Answers

1 For an answer to this question, read:

John G. Frederiksen, "Square Root +," *The Arithmetic Teacher*, November 1969, p. 549-55.

Also consult Answer 7 in Section 2.15.

2 There are two square roots of 25, 5 and −5. The symbol $\sqrt{25}$ is reserved for the positive square root, and $-\sqrt{25}$ is used for the negative square root.

## 1.14 Trigonometry

### Question

1 A teacher of eighth grade mathematics put the following figure on the board to use for drill work on trigonometric ratios.

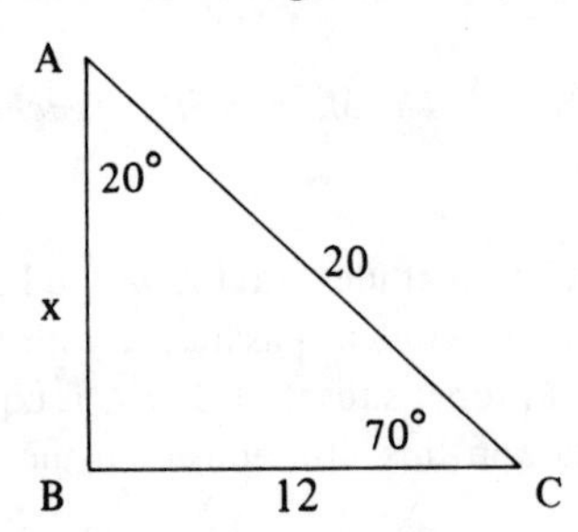


The students were told to solve for the unknown parts of △ABC. The following was discovered:

$$\tan 20° = \frac{12}{x} \qquad x = 32.6$$

$$\tan 70° = \frac{x}{12} \qquad x = 32.9$$

$$\sin 70° = \frac{x}{20} \qquad x = 18.8$$

Also by the Pythagorean theorem $M(\overline{AB}) = x = 16$. What is wrong?

## Answer

1 This problem involves an error on the part of the teacher in drawing the diagram. According to the diagram

$$\sin 20° = \frac{12}{20} = .6.$$

However, since sin 20° is approximately .342, the diagram is incorrect. A correct diagram is shown in Figure 1.14.1.

*FIGURE 1.14.1*

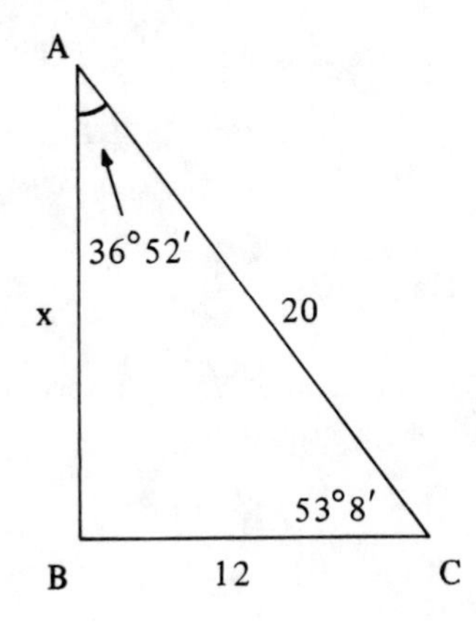

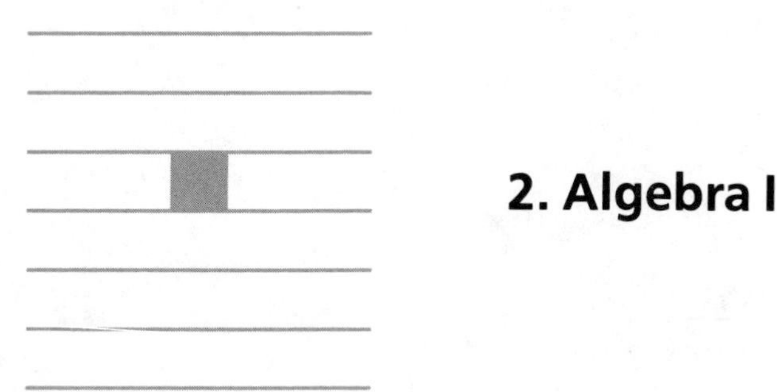

# 2. Algebra I

## 2.1 Absolute Value

### Questions

1 A student asks, "Why isn't $y = |x|$ an equation that can be represented as a line, since each term is a monomial of degree one?" How do you reply?

2 A ninth grade student gives the following definition of absolute value:

$$|x| = x \quad \text{if } x \geqslant 0.$$
$$|x| = -x \quad \text{if } x < 0.$$

Comment on her definition.

3 A student asks, "Why aren't the following equal?"

$$|x + y| = 5.$$
$$|x| + |y| = 5.$$

How do you respond?

4 How would you answer if a student asks, "What are some other uses for absolute value besides keeping distance positive?"

### Answers

1 The graph of an equation in the variables x and y is a line if and only if the equation is equivalent to an equation of the form $ax + by = c$, $a, b, c \in R$ with $a, b \neq 0$. The graph of an equation which is not equivalent to the above is not a line. Since $y = |x|$ is an equation that is not equivalent to $ax + by = c$, then the graph of this equation is not a line. If one sketches the graph of $y = |x|$ in the Cartesian plane, one gets Figure 2.1.1, and it is easy to see that this graph is not a straight line.

*FIGURE 2.1.1*

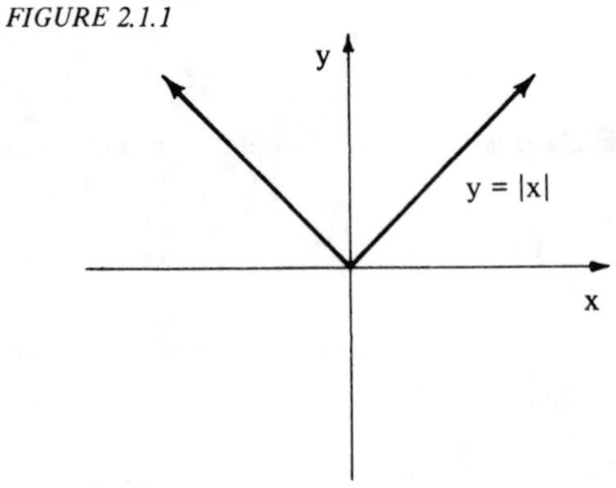


2 If $-x$ means the opposite or additive inverse of x, then her definition is correct. However, if $-x$ means negative x, then you should question what she means by this notation. If x is a negative number, then $-x$ is not usually defined since the minus sign, as used in this context, is usually part of the symbol for a negative number. It is possible that she might be using $-x$ to represent a negative number. Thus there might be a serious defect in her definition depending upon how $-x$ is interpreted.

3 $|x + y| = 5$ and $|x| + |y| = 5$ do not have the same solution sets. For example, $x = 3$ and $y = -2$ is not a solution of $|x + y| = 5$, but it is a solution of $|x| + |y| = 5$. In general, $|a + b| \leqslant |a| + |b|$, with equality only when a, b have the same sign, i.e., either both positive or negative. It might be instructive to graph both of the equations as shown in Figures 2.1.2 and 2.1.3 to give a visual demonstration of the difference.

*FIGURE 2.1.2*

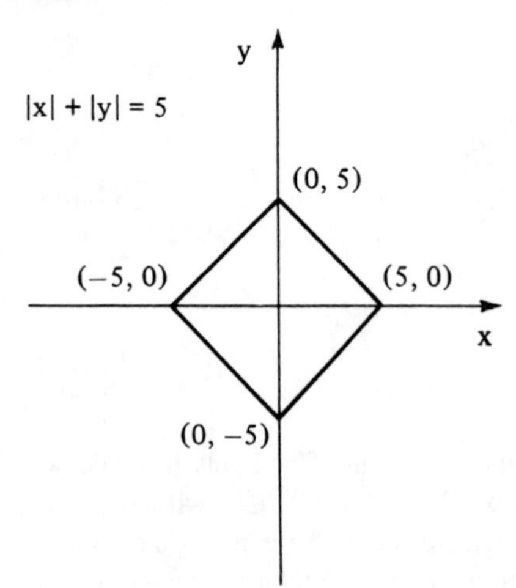


*FIGURE 2.1.3*

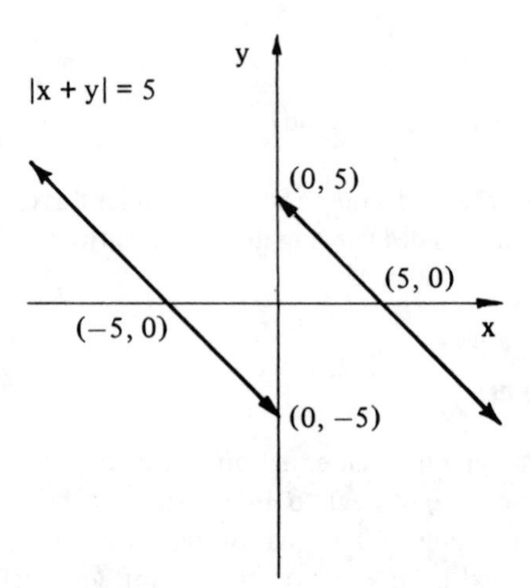


4 Refer to Section 4.1, Answer 2.

## 2.2 Cancellation

### Questions

1 While studying algebraic fractions, a student gets to this point in a simplification problem:

$$\frac{x + 5(a + b)}{a + b}.$$

The student says this is equal to $\underline{x + 5}$. How do you deal with this student's reasoning?

2 A student hands in the following work.

$$\frac{x + 6}{2} = \frac{x + \overset{3}{\cancel{6}}}{\cancel{2}} = x + 3.$$

When the student asks us why he can't cancel the 2 into the 6, how would you answer him?

3 A tenth grade student hands in the following work.

$$\frac{x^{\cancel{2}} - y^{\cancel{2}}}{x + y} = x - y.$$

The student wants to know why her method is wrong. How would you answer?

4 You ask the students to simplify $\frac{x^2 - 9}{x - 3}$. One student does the following:

$$\frac{x^2 - 9}{x - 3} = \frac{\cancel{x}^{\cancel{2}} - 9}{\cancel{x} - 3} = \frac{x - 9}{-3} = \frac{x + 9}{+3} = \frac{x + \overset{3}{\cancel{9}}}{\cancel{3}} = x + 3.$$

Is the student correct? Will his method work all the time?

5 You ask the students to simplify $\frac{1}{x} + \frac{1}{x}$. A student writes

$$\frac{1}{x} + \frac{1}{x} = 2.$$

What is this student's problem?

6 A student asks you if $\sqrt{a^2 + b^2} = a + b$? How do you respond?

7 A student is evaluating the expression

$$\frac{3^2 + 7^2}{\left(\frac{58}{10}\right)^2},$$

and the student does the following:

$$\frac{3^{\not 2} + 7^{\not 2}}{\left(\frac{58}{10}\right)^{\not 2}} = \frac{10}{\left(\frac{58}{10}\right)} = \frac{100}{58}.$$

Which is the correct answer? How would you help this student?

8 A student is asked to simplify the following:

$$\frac{x + \frac{1}{x}}{x - \frac{1}{x}}.$$

His work is: $$\frac{\not x + \frac{\not 1}{\not x}}{\not x - \frac{\not 1}{\not x}} = \frac{1}{-1} = -1.$$ How would you help this student?

## Answers

1,2 The type of errors made in Questions 1 and 2 occur frequently in an Algebra I class and is one of the reasons why many teachers do not like to use the term *cancel* when teaching this course. One way of handling this situation is to make sure that students realize at the beginning of the course that *cancelling* really involves the identity element for multiplication, i.e., multiplying by 1. Thus many problems of this nature should be worked out in great detail by the teacher and the class. The teacher should also point out the difference between problems such as Questions 1 and 2, where "cancelling" is done incorrectly, with the following problems, which are correctly simplified.

(a) $$\frac{x \cdot 5(a + b)}{a + b} = 5x \cdot \left(\frac{a + b}{a + b}\right) = 5x$$

(b) $$\frac{x \cdot 6}{2} = 3x \cdot \left(\frac{2}{2}\right) = 3x$$

The teacher might point out that the answer, $x + 5$, is the answer to the problem of simplifying

$$\frac{x(a + b) + 5(a + b)}{a + b} = \frac{(x + 5)(a + b)}{a + b} = x + 5.$$

For Question 2 the teacher might point out that the answer, $x + 3$, is the answer to the problem of simplifying

$$\frac{2x + 6}{2} = \frac{2(x + 3)}{2} = \frac{2}{2} \cdot \frac{x + 3}{1} = x + 3.$$

However, there does not seem to be a sure-fire method which will always guarantee success in this situation. It seems to be the case that at least one student will make at least one of these errors at some point in an algebra course.

3 If the student's method was correct, then she should be able to get the correct answer on all problems of this type. Thus, you might ask her to do the following problem.

$$\frac{x^2 + y^2}{x + y}$$

Using her method the student would get x + y, but

$$(x + y)(x + y) \neq x^2 + y^2.$$

For example, take x = y = 1, then one might proceed to show the student a correct method; i.e.,

$$\frac{x^2 - y^2}{x + y} = \frac{(x + y)(x - y)}{x + y} = \frac{x + y}{x + y} \cdot (x - y) = x - y,$$

assuming, of course, that $x + y \neq 0$.

4 Even though the student gets the correct answer in this case, the method will not work all the time. Consider the following problem.

$$\text{Simplify } \frac{x^2 + 9}{x + 3}.$$

Using the student's method, we get

$$\frac{x^2 + 9}{x + 3} = \frac{x^{\not 2} + 9}{\not x + 3} = \frac{x + 9}{+3} = \frac{x + \not 9^{3}}{\not 3} = x + 3.$$

But, $(x + 3)(x + 3) \neq x^2 + 9$ unless x = 0.

5 It is very difficult to pinpoint exactly what the student's problem is, without talking further to the student, but two possibilities, which seem to be good bets, are given below.

(a) One possibility is that the student cancelled incorrectly as follows:

$$\frac{1}{\underset{1}{\not x}} + \frac{1}{\underset{1}{\not x}} = \frac{1}{1} + \frac{1}{1} = 2.$$

If this is the case, then you might try to show the student that the answer is incorrect by substituting values in for x; i.e., if x = 4, then

$$\frac{1}{x} + \frac{1}{x} = \frac{1}{4} + \frac{1}{4} = \frac{1}{2}.$$

(b) Another possibility is that the student thought along the following lines. "If I have one object plus another object, then altogether I have *two* objects." Then the student just wrote *two* for the answer without attaching the label (i.e., $2 \cdot \frac{1}{x}$).

6 No, $\sqrt{a^2 + b^2} \neq a + b$. One way to see this is to consider $a = -5$ and $b = 5$. Then

$$\sqrt{a^2 + b^2} = \sqrt{(-5)^2 + (5)^2} = \sqrt{50} = 2\sqrt{5}.$$

However, $a + b = -5 + 5 = 0$. Since $2\sqrt{5} \neq 0$, then $\sqrt{a^2 + b^2} \neq a + b$.

Another way of looking at this situation is to assume that $\sqrt{a^2 + b^2} = a + b$. Then

$$a^2 + b^2 = (a + b)^2 = a^2 + 2ab + b^2.$$

However, for a and $b \neq 0$, $a^2 + b^2 \neq a^2 + 2ab + b^2$.

7 One way of helping this student is to find a counterexample such as

$$\frac{3^2 + 4^2}{5^2} = \frac{25}{25} = 1.$$

Using the student's method, one gets

$$\frac{3^2 + 4^2}{5^2} = \frac{3^{\not{2}} + 4^{\not{2}}}{5^{\not{2}}} = \frac{7}{5}.$$

The student's method just happened to work on this particular exercise. An interesting problem would be to find all such problems like the one above for which the student's method gives the correct solution; i.e., what are the conditions on a, b, c, and d such that

$$\frac{a^2 + b^2}{\left(\frac{c}{d}\right)^2} = \frac{a + b}{\frac{c}{d}}\ ?$$

8 One way of helping this student would be as follows. The student's work indicates that the expression on the left is the same as the expression on the right for all defined values of x. Suppose we substitute 2 for x in the expression on the left.

$$\frac{x + \frac{1}{x}}{x - \frac{1}{x}} = \frac{2 + \frac{1}{2}}{2 - \frac{1}{2}} = \frac{\frac{5}{2}}{\frac{3}{2}} = \frac{5}{3}.$$

Thus it would seem that something is wrong. The main problem here is that the student has confused the processes of simplifying the following expressions:

(a) $\dfrac{x + \frac{1}{x}}{x - \frac{1}{x}}$ and (b) $\dfrac{x \cdot \frac{1}{x}}{x \cdot \frac{-1}{x}}$

What the student has done is to use a process which would correctly simplify (b). Notice that in (b) one has multiplication and the expression involves factors, but in (a) there is addition and the expression involves addends.

## 2.3 Decimals and Percent

### Questions

1 A student who was told that .999... = 1 says, "If .999... = 1, then you should be able to change $\frac{9}{9}$ to .999... by dividing the denominator into the numerator. But $9\sqrt{9.00} = 1.00$ not .999..." Can you show the student how $9\sqrt{9.00}$ equals .999... and explain why you can do that?

2 A student hands in the following work.

$$n\% = \frac{n}{100} = 0.0n.$$

Is the student correct?

### Answers

1 When presenting the algorithm of division, it is usually standard to present the following assertion.

> For given integers a and b with $b > 0$, there exist unique integers q and r such that $a = b \cdot q + r, 0 \leqslant r < b$.

Then, the practice is to show how this assertion is related to our usual algorithm for division. However, one can modify this assertion if one wishes, although it is not standard practice, so that the restrictions, *unique integers* and $0 \leqslant r < b$, are omitted. If this is done, then one could answer the student's question by doing the following:

```
    .999...
9 |9.000...
   8.1
    90
    81
    90
```

At this level a student probably would not understand the subtleties involved mathematically, but the student could be expected to follow the arithmetic.

2 No, the student is not correct. Consider n = 12, then 12% is equal to 12/100, but $12/100 \neq 0.012$. Note that the student is correct if *n* is a digit.

## 2.4 Equations, Solving Equations, Equivalence

### Questions

1 While solving the equation,

$$\frac{x}{x-1} = 1 + \frac{1}{x-1},$$

a student gets down to the part where $x = x$ and $x \neq 1$. The student says that he doesn't know where to go from here. How would you help this student?

2 A student solves $\frac{3}{8} = \frac{x+2}{16}$ in the following way.

$$\frac{3-2}{8-2} = \frac{x+2-2}{16-2}$$

$$\frac{1}{6} = \frac{x}{14}$$

$$6x = 14$$

$$x = \frac{14}{6}.$$

Is the student's reasoning correct?

3 A student hands in the following solution to this problem. "Solve the following equation by factoring: $4x^2 - 23x + 15 = 0$."

$$4x^2 - 23x + 15 = 0$$

$$\dot{x}^2 - 23\dot{x} + 60 = 0$$

$$(\dot{x} - 20)(\dot{x} - 3) = 0$$

$$\dot{x} = 20 \quad \text{or} \quad \dot{x} = 3.$$

Therefore $x = 20/4 = 5$ or $x = 3/4$. Is this method correct?

4 A student asks you if every equation of a line is a first degree equation in x and y. What is your response?

5 You are asked: "What is the difference between an open phrase and an open equation?" What do you tell this student?

6 You ask your students to solve the equation $x/7 = 3$. One of your students raises his hand and says that it is impossible to do this problem because the domain of the variable has not been specified. What is your response?

7 A student asks you if $x/x$ and $(x - 3)/(x - 3)$ are equivalent expressions. What is your response?

8 The students in your class are asked to solve the following problem. "Find the roots of $x^3 = 2x^2 + 8x$." One student does the following work.

$$x^3 = 2x^2 + 8x$$
$$x^2 - 2x - 8 = 0$$
$$(x - 4)(x + 2) = 0$$
$$x = 4 \text{ or } x = -2$$
$$x \in \{4, -2\}.$$

Is the student correct?

## Answers

1 The logic involved in solving open sentences is to use properties of the algebraic system to transform an open sentence into an equivalent open sentence; that is, the open sentences have the same solution set. Thus when the student gets to the part where $x = x$ and $x \neq 1$, the student should realize that the solution set of the original open sentence is equivalent to the solution set of this open sentence. By inspection the solution set of $x = x$ and $x \neq 1$ is the set of all numbers of the domain except 1. Thus the solution set of $\frac{x}{x-1} = 1 + \frac{1}{x-1}$ is the set of all numbers of the domain except 1.

2 No, the student's reasoning is not correct. Solving the equation in the standard way, one gets {4} as the solution set. The student is trying to use a procedure that he probably made up (as students often do), which is not mathematically correct. The student is assuming that

$$\frac{a}{b} = \frac{c}{d} \text{ implies } \frac{a-k}{b-k} = \frac{c-k}{d-k}$$

where k is a real number. That this is not correct can be shown by taking $a = 1$, $b = 2$, $c = 3$ and $d = 6$. With $k = 1$, one obtains the false statement $0 = 2/5$.

***Problem*** Under what conditions will the student's method work?

3 Yes, this method is correct as illustrated here. Let $ax^2 + bx + c = 0$, $a \neq 0$ be the expression for the general quadratic equation in the variable x. Using the quadratic formula, one gets the following solutions.

$$x = \frac{-b \pm \sqrt{b^2 - 4ac}}{2a} \tag{1}$$

Using the student's method, let $\dot{x}^2 + b\dot{x} + ac = 0$ be the general expression in the variable $\dot{x}$. Using the quadratic formula, one gets

$$\dot{x} = \frac{-b \pm \sqrt{b^2 - 4ac}}{2}. \tag{2}$$

If one compares (1) and (2), then it can be seen that if (2) is divided by *a*, then the result is (1).

4 This question seems to have at least two different interpretations depending on what the student is really asking. In answering this question, the assumption is made that the student is referring to lines in two dimensions.

If the student is asking if the *only* way to write an equation of a line is as a first degree equation in x and y, then the answer is no since there are different ways to write equations for lines such as parametric and vector equations.

If the student is asking if every *rectangular* equation of a line is a first degree equation in x and y, then the answer depends on how a first degree equation in x and y is defined. If in (1) below neither a nor b can be zero, then the answer is no because of the special cases $y = k$ and $x = f$, where k and f are real numbers. If a or b, but not both, can be zero, then the answer is yes.

$$ax + by = c \qquad a, b, c \text{ real numbers} \tag{1}$$

5 An open equation is an open sentence such as $3x + 2 = 4$ or $3x = 1$. An open phrase is an expression such as $3x + 2$ or $3x$. Thus an open phrase is an algebraic expression that contains one or more variables but no equality sign, inequality sign, etc. The situation is similar to the difference between a phrase and a sentence in English. A sentence contains a verb whereas a phrase does not.

6 Technically speaking, the student is correct. The domain of the variable should always be specified. However, usually the convention is made that if the domain is not specified, then one should assume that the domain consists of all those elements of the largest number system we are currently working with for which the expression is defined.

7 By definition two algebraic expressions are equivalent if and only if the values of the expressions are equal for every numerical replacement of the variables. Thus, for every value of the variables, equivalent expressions represent the same number. Thus the expression $x/x$ and $(x - 3)/(x - 3)$ are not equivalent since when $x = 3$, $x/x = 1$, while $(x - 3)/(x - 3)$ is undefined. Likewise when $x = 0$, $x/x$ is undefined while $(x - 3)/(x - 3) = 1$.

Another, but similar, way to respond is to consider $x/x$ and $(x - 3)/(x - 3)$ as functions. The domain of $x/x$ would be all real numbers except 0 while the domain of $(x - 3)/(x - 3)$ would be all real numbers except 3. Since the domains differ, the functions differ and hence $x/x$ and $(x - 3)/(x - 3)$ are not equivalent expressions.

8 No, the student is not correct. In the second step, the student divided through by x, and this is valid as long as $x \neq 0$. The student must then find out what happens when $x = 0$. Is it a solution or not? In this particular case, 0 is a solution; thus the solution set should be $\{0, -2, 4\}$.

## 2.5 Exponents

### Questions

1 In your ninth grade class, a student hands in the following solution. "$(a + b)^2 = a^2 + b^2$ because, except for the operation being different, this problem is just like $(a \cdot b)^2$, which is equal to $a^2 \cdot b^2$. Therefore, it follows that $(a + b)^2 = a^2 + b^2$." How would you help this student?

2 A student asks, "Can you change a positive exponent to a negative one by taking the reciprocal of the base? Does $2^{2/3} = (1/2)^{-2/3}$?" How do you reply?

3 A student asks, "Why isn't $2^5 \cdot 2^{-5} = 4^0 = 1$?" How do you reply?

4 A student asks you if multiples of a number under addition are related to powers of a number under multiplication. How do you respond?

5 You are asked: "How would you evaluate the expression $2^{3^2}$? Is it $(8)^2 = 64$ or $2^9 = 512$?" How would you reply?

6 You ask the class whether they would rather have your annual salary of $10,000 or would they like to have the total amount of money they would get in the following way: one cent on the first day, two cents on the second day, four cents on the third day, eight cents on the fourth day, etc. for thirty days. The class chooses your salary, and then you proceed to tell them that from a computer you determined that they would have received $10,737,400 by the other method. However, one student says that the computer is wrong and shows you the following work. "We would have received $2^{30}$ cents. Now, $2^5 = 32$, and therefore $(2^5)^2 = \$10.24$. Thus $2^{30} = (10.24)^3$, which is about $1004.86." Who is correct, the computer or the student?

### Answers

1 Although the student should be commended for looking for patterns in mathematics, one must be wary of extensions and this particular pattern does not extend to addition. Several different ways of explaining this to a student are given below.

(a) If $(a + b)^2 = a^2 + b^2$, then this equality should hold for all values of a and b. Let $a = 2$ and $b = 3$, then $(a + b)^2 = (2 + 3)^2 = 5^2 = 25$, while $a^2 + b^2 = 2^2 + 3^2 = 13$ and $25 \neq 13$.

(b) A geometric consideration of $(a + b)^2$ could help students. $(a + b)^2$ is the area of the square shown in Figure 2.5.1 and it is easily seen that

$$(a + b)^2 = a^2 + b \cdot a + a \cdot b + b^2$$
$$\therefore (a + b)^2 = a^2 + 2ab + b^2$$

Thus, in general, $(a + b)^2 \neq a^2 + b^2$.

*FIGURE 2.5.1*

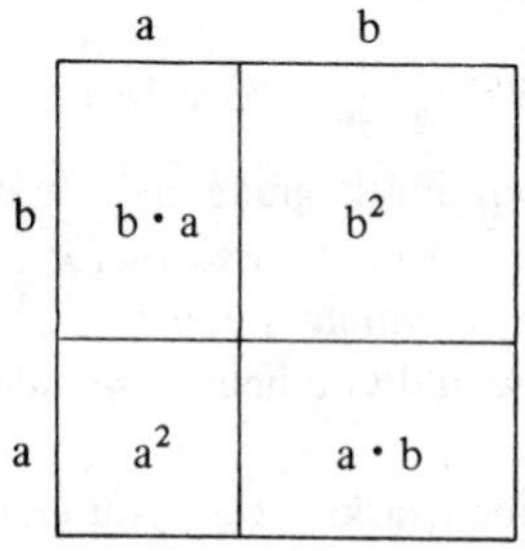

| | a | b |
|---|---|---|
| b | $b \cdot a$ | $b^2$ |
| a | $a^2$ | $a \cdot b$ |

2 Yes. Let $a$ be a real number not equal to zero, and let r and s be real numbers, $s \neq 0$. Then $(a)^{r/s} = (a^{-1})^{-r/s}$.

3 In this particular case, the student gets the correct answer. However, it seems as if the student is using an incorrect method. For example, if the student is asked to simplify $2^5 \cdot 2^2$, it looks as if the student would get $4^7$ instead of $2^7$.

4 Yes, they are isomorphic systems. A few of the properties are illustrated as follows. Let $a \neq 0$ be a real number. Then

$$ma \longleftrightarrow a^m$$
*addition* *multiplication*

is an isomorphism and every property of multiplies has a corresponding property of powers.

(a) The multiple of a multiple is a multiple. The power of a power is a power.

$n(ma) = (n \cdot m)a$ $(a^m)^n = a^{mn}$

(b) The multiple of a sum is the sum of multiples. The power of a product is the product of the powers.

$m(a + b) = ma + mb$ $(ab)^m = a^m b^m$

5 It is conventional in mathematics to evaluate the expression $2^{3^2}$ as $2^{(3^2)}$ or $2^9$, i.e., from the top to the bottom. Notice that this example says that the operation of raising to a power is not associative. For an interesting article concerning this problem when there are an infinite number of terms, consult:

M.C. Mitchelmore, "A Matter of Definition," *The Mathematical Monthly*, (Classroom Notes Section), June-July 1974, pp. 643-7;

***Problem*** Find all values of x, y, z such that $(x^y)^z = x^{(y^z)}$.

6 The computer is correct and the student is wrong. The following computer program is given which shows the amount received each day in the first column and the total sum of money received in the second column. Thus at the end of thirty days, the student would have received $10,737,400.

***Problem*** Why is the student incorrect?

```
LIST
MONEY    03:06 PM           50 S=S+Z
10 S=0                      51PRINT Z,S
20 FORI=0TO 63              60 NEXT I
30 LET Z = 2↑I/100          80 END

       READY

       RUNNH
 1      .01                  .01
 2      .02                  .03
 3      .04                  .07
 4      .08                  .15
 5      .16                  .31
 6      .32                  .63
 7      .64                  1.27
 8      1.28                 2.55
 9      2.56                 5.11
10      5.12                 10.23
        10.24                20.47
        20.48                40.95
        40.96                81.91
        81.92                163.83
        163.84               327.67
        327.68               655.35
        655.36               1310.71
        1310.72              2621.43
        2621.44              5242.87
20      5242.88              10485.7
        10485.8              20971.5
        20971.5              41943
        41943                83886.1
        83886.1              167772
        167772               335544
        335544               671089
        671089               .134218E 7
        .134218E 7           .268435E 7
        .268435E 7           .536871E 7
30      .536871E 7           .107374E 8
        .107374E 8           .214748E 8
```

## 2.6 Factoring and Factors

### Questions

1 A student says that 3a − 2a = 1. How would you help him?

2 A student is trying to factor $y^2 - 2xy + x^2 - w^2$. The student's work is as follows:

$$(y - x)^2 - w^2 = (y - x - w)(y - x + w).$$

However, another student rewrites the problem as $x^2 - 2xy + y^2 - w^2$, and the work is as follows:

$$(x - y)^2 - w^2 = (x - y - w)(x - y + w).$$

Which way is correct?

3 You are teaching factoring of quadratic polynomials. A student asks, "Is there any other way, other than trial and error?" How would you answer this student?

4 You have given your students a quiz on factoring polynomials. After grading the quiz and handing it back to the class, a number of students complain that you should not have marked (c) wrong in the following problem.

Factor the polynomials listed below.

(a) $x^2 - y^2$ (b) $9y^2 - 25x^2$ (c) $3x + 8$

The work in (c) is: $3x + 8 = 3\left(x + \frac{8}{3}\right)$. What is your reply.

5 A student asks you if zero is considered a factor of any number. What is your reply?

6 A student hands in the following work for the following problem.

"Solve: $x^2 - 14x + 24 = 3$."

$$(x - 12)(x - 2) = 3$$
$$(x - 12)(x - 2) = 3 \cdot 1$$
$$x - 12 = 3 \quad \text{or} \quad x - 2 = 1$$
$$x = 15 \qquad x = 3$$
$$x \in \{3, 15\}.$$

Is the student correct?

7 A student asks, "Is it possible to factor a trinomial (when it can be factored) by an exact method rather than the usual trial and error method?" How would you respond?

8 A student asks, "Why is this procedure called *completing the square*; a square is a geometric shape?" How do you reply?

## Answers

1 This error by a student is one that frequently comes up in an algebra course, and there are various ways that one may help the student. One way would be to use an inductive approach as shown in Table 2.6.1.

*TABLE 2.6.1*

| *Value of a* | *Value of 3a – 2a* |
|---|---|
| 1 | $3 - 2 = 1$ |
| 2 | $6 - 4 = 2$ |
| 3 | $9 - 6 = 3$ |
| 4 | $12 - 8 = 4$ |
| 5 | $15 - 10 = 5$ |
| 6 | $18 - 12 = 6$ |

The student should now realize that $3a - 2a$ seems to be always equal to $a$. Then you might ask the student if he can find a value of $a$ for which this would not be true. After the student tries a few more cases, then you might want to show him that $3a - 2a = a$ by using the distributive property of multiplication over subtraction as follows.

$$3a - 2a = (3-2) \cdot a = 1 \cdot a = a$$

2 Both students are correct.

$$\begin{aligned}(y - x - w)(y - x + w) &= (-x + y - w)(-x + y + w)\\ &= [-(x - y + w)]\,[-(x - y - w)]\\ &= (x - y + w)(x - y - w)\end{aligned}$$

Therefore, $y^2 - 2xy + x^2 - w^2 = (y - x - w)(y - x + w) = (x - y + w)(x - y - w)$

3 For an answer to this question, read the following article by G.H. Palagi.*

# A CONVERSATION ON FACTORING

**By GEORGE H. PALAGI**

**Lincoln Junior High School**
**Billings, Montana**

STUDENT: When the coefficient of the quadratic term in a trinomial is not 1, I sometimes have a lot of trouble finding the factors.

PROFESSOR: Yes, the number of combinations to be tried on a trial and error basis can be quite large.

STUDENT: I don't suppose there is a "general form," of which you are so fond, that can be applied to factoring trinomials, or is there?

PROFESSOR: Let us see if we are able to find one. Perhaps if we work the problem backwards, so to speak, we will at least find a general approach to the problem.

STUDENT: I should be of help in that case for I quite often get things backwards.

PROFESSOR: Come now, you have a nice feeling for mathematics. Suppose we had worked the problem and this was our answer:

$$(ax + b)(cx + d).$$

Apply the distributive axiom and what do we get?

*From *The Mathematics Teacher*, November 1973, pp. 671-672. Reprinted by permission.

STUDENT: I suppose we would get

$$(ax + b)cx + (ax + b)d.$$

PROFESSOR: Good. Now apply the "right" distributive axiom twice.

STUDENT: Then we get

$$acx^2 + bcx + adx + bd.$$

PROFESSOR: Now apply the same axiom to the linear terms.

STUDENT: Let's see. We get

$$acx^2 + (bc + ad)x + bd.$$

PROFESSOR: Excellent. Now by studying this result, we shall, perhaps, be able to make some progress towards finding a general solution, if there is one.

STUDENT: This is the part I usually don't see.

PROFESSOR: Let us study a little more and ask ourselves some questions. If we were to multiply the coefficient of the quadratic term by the constant term, we would have

$$abcd.$$

STUDENT: And the coefficient of the linear term, if multiplied instead of added, is also

$$abcd.$$

PROFESSOR: That should give us a clue. Let us see if we can apply this to a specific problem, for instance,

$$6x^2 + 19x + 15.$$

STUDENT: The product of the quadratic term and the constant is $90x^2$. So we want to replace the coefficient of the linear term, $19x$, with the factors of $90x^2$ that add up to $19x$.

PROFESSOR: Precisely, and in this case it is $9x + 10x$.

STUDENT: Then we have

$$6x^2 + 9x + 10x + 15.$$

PROFESSOR: Apply the "left" distributive axiom to the first two terms and also to the last two terms.

STUDENT: Then we have

$$3x(2x + 3) + 5(2x + 3).$$

PROFESSOR: Now use the "right" distributive property.

STUDENT: Now we have

$$(3x + 5)(2x + 3).$$

PROFESSOR: And so our algorithm works on this problem.

STUDENT: Will it work on this one?

$$15x^2 - 7x - 2.$$

PROFESSOR: Of course. But let us work it out. The product of the quadratic term and the constant term is $-30x^2$; so we want to replace $-7x$ with the factors of $-30x^2$ that add up to $-7x$.

STUDENT: And they are $-10x$ and $+3x$.

PROFESSOR: So now we have

$$15x^2 - 10x + 3x - 2$$

$$5x(3x - 2) + (3x - 2)$$

$$(5x + 1)(3x - 2).$$

STUDENT: Does it matter if we put $3x - 10x$ rather than $-10x + 3x$?

PROFESSOR: No, for when we were working in the general form, we could apply the commutative axiom of addition to $(bc + ad)x$ and get $(ad + bc)x$. But let us work it out anyway.

$$15x^2 + 3x - 10x - 2$$

$$3x(5x + 1) - 2(5x + 1)$$

$$(3x - 2)(5x + 1).$$

PROFESSOR: Once we factor the first two terms, we must be careful to get a common factor when we factor the third and fourth terms.

STUDENT: Suppose the trinomial is prime? What happens then?

PROFESSOR: Do you mean by prime a polynomial that cannot be factored into polynomials of lower positive degree with integral coefficients and whose greatest common monomial factor is 1?

STUDENT: Yes, that is what I meant.

PROFESSOR: Let us work a specific problem and then see if we can generalize from that.

STUDENT: How about

$$4x^2 + 2x - 1?$$

PROFESSOR: Fine. The product of the quadratic term and the constant term is $-4x^2$. The possible pairs are

$$-4x,\ x$$
$$-2x,\ 2x$$
$$-x,\ 4x$$
$$4x,\ -x$$
$$2x,\ -2x$$
$$x,\ -4x$$

STUDENT: And none of those add up to $+2x$ and so we know it is prime.

PROFESSOR: And now, just to make sure you can use this algorithm, work this problem,

$$10x^2 - 11x - 6.$$

STUDENT: Let's see. The product of the constant and the quadratic term is $-60x^2$. That means we have these possible pairs,

| | |
|---|---|
| $x,\ -60x$ | $-x,\ 60x$ |
| $2x,\ -30x$ | $-2x,\ 30x$ |
| $3x,\ -20x$ | $-3x,\ 20x$ |
| $4x,\ -15x$ | $-4x,\ 15x$ |
| $5x,\ -12x$ | $-5x,\ 12x$ |
| $6x,\ -10x$ | $-6x,\ 10x$ |

And the pair that adds up to $-11x$ is $4x$ and $-15x$. Then we have

$$10x^2 + 4x - 15x - 6$$
$$2x(5x + 2) - 3(5x + 2)$$
$$(2x - 3)(5x + 2).$$

I think I've got it now. Thank you, Professor, for your help.

PROFESSOR: I'm always glad to help when I can.[1]

1. The reader who is interested in a more complete exposition is referred to Drooyan, Hadel, and Fleming, *Elementary Algebra, Structure and Skills* (New York: John Wiley and Sons, 1966, 1969).

4 The assumption here is that the factoring is done over the set of integers. If this was specificed beforehand, then the students were wrong. However, if this was not specified beforehand, then the students are correct. They have factored $3x + 8$ over the set of rational numbers.

5 We say that *a* is a factor of c over a given set of numbers if and only if $a \cdot b = c$ where a, b, c belong to the given set. Thus 0 is a factor of 0 over a set of numbers because $0 = 0 \cdot r$ where r could be any number from the set of given numbers. Thus 0 is a factor of only one number, 0, because $0 \cdot b = c$ would only be true when c is zero.

6 By substituting 3 and then 15 into the given equation, we get

| | | |
|---|---|---|
| $3^2 - 14 \cdot 3 + 24 = 3$ | and | $15^2 - 14 \cdot 15 + 24 = 3$ |
| which simplifies to | | which simplifies to |
| $-9 = 3$, | | $39 = 3$, |
| a *false* statement | | a *false* statement |

Therefore, the student is not correct. The student is trying to use the theorem that "if a and b are real numbers and a • b = 0, then a = 0 or b = 0." However, an essential condition is that the product must be equal to zero. One cannot generalize this theorem for the situation where a product is equal to a nonzero number.

***Problem*** Under what conditions will the student's method work?

7 See Answer 3 of this section.

8 Suppose we use the equation $x^2 + 9x - 36 = 0$ as an example. Now,

$$x^2 + 9x - 36 = 0$$

can be written as $x^2 + 9x = 36$. Geometrically we have the situation shown in Figure 2.6.1.

*FIGURE 2.6.1*

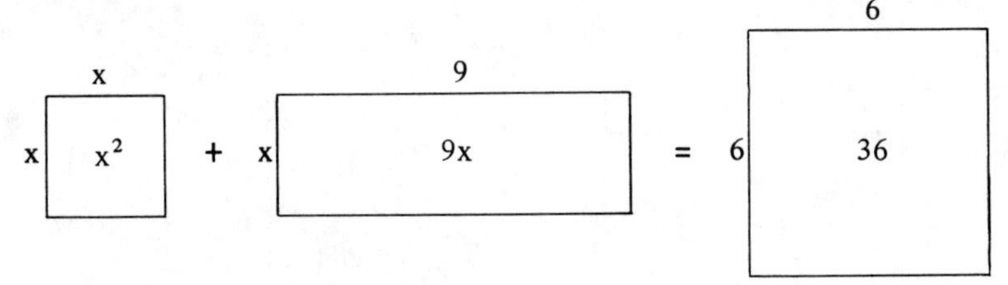


Let us now divide the second rectangle in half as in Figure 2.6.2.

*FIGURE 2.6.2*

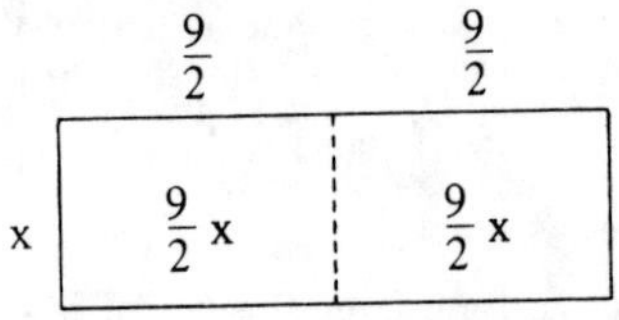

Then we attach each half to the square of side x as in Figure 2.6.3.

*FIGURE 2.6.3*

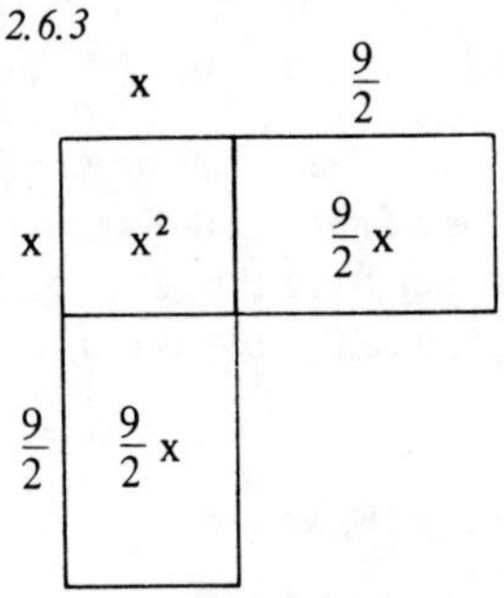

In order to complete the square, we must add a square of side 9/2 as shown in Figure 2.6.4.

*FIGURE 2.6.4*

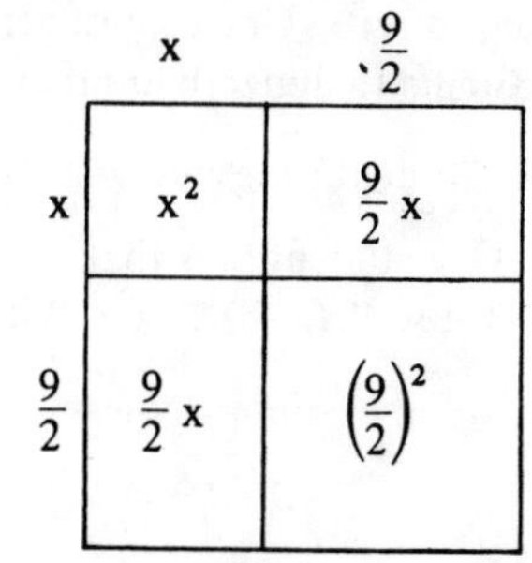

Geometrically, we get the situation shown in Figure 2.6.5.

*FIGURE 2.6.5*

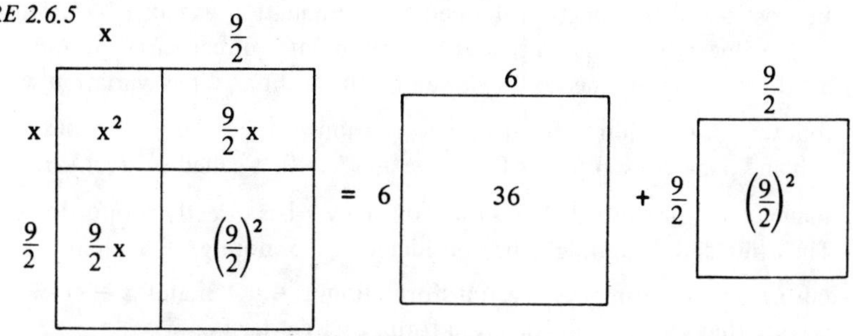

Algebraically we arrive at the following solution.

$$x^2 + 9x = 36$$

$$x^2 + 9x + \left(\frac{9}{2}\right)^2 = 36 + \left(\frac{9}{2}\right)^2$$

$$\left(x + \frac{9}{2}\right)^2 = \frac{225}{4}.$$

Thus this process of solving a quadratic equation gets its name *completing the square* from a geometric process.

## 2.7 Functions and Relations

### Questions

1 A student asks you if all linear functions are direct variations. What is your response?

2 A student suddenly smiles and makes the following assertion: "Addition, subtraction, multiplication, and division are all examples of functions, aren't they?" What response do you give?

3 A student says, "The function $f(x) = 3x + 2$ is a linear function." Do you agree?

4 You are asked if equation and function mean the same thing. What is your reply?

5 You ask your class the following question. "What is the relation between $A \times B$ and $B \times A$ when A and B are arbitrary sets?" One student says that it is an illustration of the symmetric property of equality because if $(a, b) \in A \times B$, then $(b, a) \in B \times A$. Do you agree?

## Answers

1 In most texts, the function defined by the equation $y = ax + b$ is normally considered a linear function, but does not represent a direct variation unless $b = 0$ and $a \neq 0$. However, in one text reviewed by the authors, direct variation was defined as follows. A function defined by an equation of the form $y + b = k(x + a)^n$ when k, a, b and n are constants, $k \neq 0$, $x \neq -a$ and $n > 0$, is called a direct variation. We say that $y + b$ varies directly as $(x + a)^n$ or that $y + b$ is directly proportional to $(x + a)^n$. Thus this definition might be considered to be more general than the one given above. But if one makes the transformation $Y = y + b$ and $X = (x + a)^n$, then one can see that they are equivalent in form, i.e., $Y = k \cdot X$.

2 You might ask the student to explain what she means by that statement. The student's explanation should be somewhat along the following lines. For every ordered pair of numbers where defined, each operation assigns only one number. Thus each operation, over the appropriate domain, is a mapping of $A \times A \rightarrow A$ and is therefore a function. The domain in each case is a set of ordered pairs while the range is set A. For example, A is the set of real numbers, then one can regard addition as a function

$$+ : A \times A \rightarrow A$$

with $+(2, 6) = 8$; $+(3, 1) = 4$, etc.

3 The problem here is one of long standing and involves precise use of language. Technically speaking the student is incorrect. The expression f(x) refers to an element in the range of the function, not the function itself. Strictly speaking, x refers to an element in the domain of a function, f(x) refers to the element in the range corresponding to the element x in the domain and f refers to the function itself. However, it is often the case, especially in advanced mathematics, where f and f(x) are used interchangeably with the context of the instruction showing which interpretation is intended.

For more information on this subject, read:

(1) R.B. Deal, "Some Remarks on 'A Dilemma in Definition'," *The Mathematical Monthly*, Vol. 74, No. 10, p. 1258.

(2) H.A. Thurston, "A Reply to 'A Dilemma in Definition'," *The Mathematical Monthly*, Vol. 74, No. 10, p. 1259.

4 Equations and functions are not the same thing. An equation often, although not always, determines a function. Thus $y = 2x + 1$ is an equation that describes the rule used to determine a function and as such is not a function itself. Again in various situations, such as in advanced mathematics, impreciseness is tolerated in that both terms are used interchangeably. This convention is done in order to simplify the writing chore in mathematics texts.

Another difficulty is that an equation such as $x^2 + y^2 = 1$ does not give x as a (single-valued) function of y nor y as a function of x.

5 Although the student shows good thinking, he is incorrect. It seems as if he means that if $(a, b) \in A \times B$, then $a = b$. However, it does not follow that the symmetric property of equality is the only relation illustrated in this case; i.e., that $(a, b) \in A \times B$ infers $a = b$ and $(b, a) \in B \times A$ infers $b = a$. One should probably question the student in order to find out exactly what the student means. The correct answer should be that $A \times B$ and $B \times A$ are equivalent sets (i.e., they contain the same number of elements).

## 2.8 Graphs

### Questions

1 You give your students the following problem. Graph $x = 3$. One student hands in the following work.

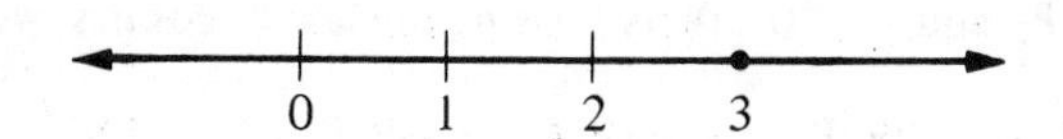


Another student hands in the work below.

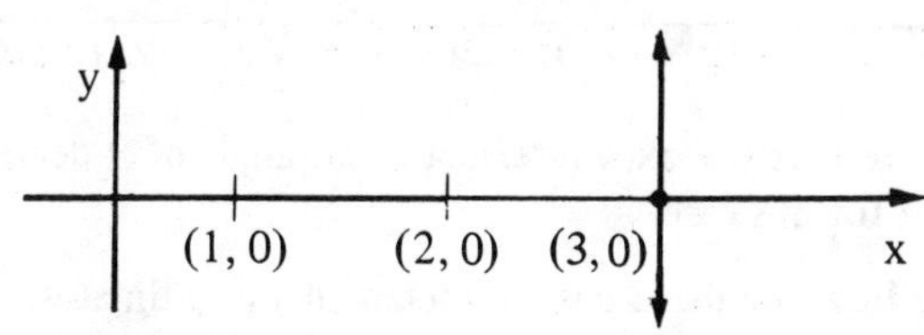


Which student is correct?

2 You are teaching the distance formula to your class, and a student asks, "Suppose that the coordinate axes are not perpendicular but they form some other angle (i.e., oblique axes), say, 60°. How would you find the distance between points in this system?" How would you respond?

3 While studying a unit on the graphs of straight lines, a student asks you if having the slope zero means the same thing as having no slope. What is your response?

## Answers

1 Since a restriction was not included as to whether the students should graph in one or two-dimensional spaces, both students are correct. Marking one of the students correct and the other one incorrect would be quite unfair.

2 Suppose that $P_1(x_1, y_1)$ and $P_2(x_2, y_2)$ are two points in this coordinate system as shown in Figure 2.8.1.

*FIGURE 2.8.1*

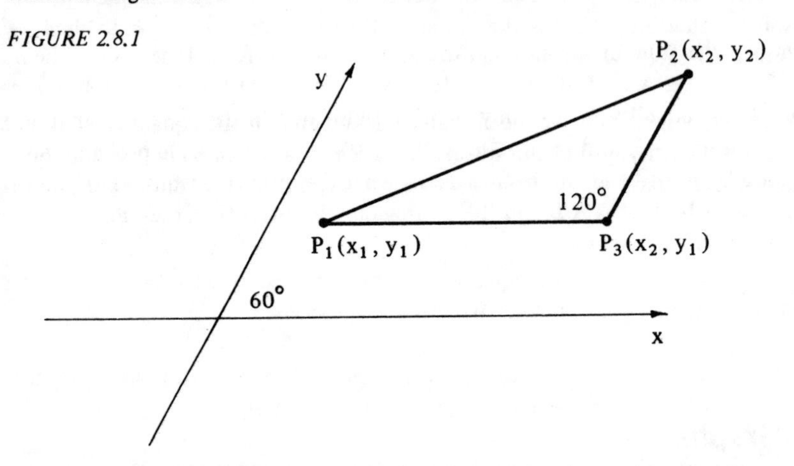


In order to find the distance between $P_1$ and $P_2$, we proceed in the usual case for rectangular coordinates by drawing segments parallel to the x and y axes. Let $P_3$ be the point of intersection of these two segments. The coordinates of $P_3$ are $(x_2, y_1)$ and angle $P_1P_3P_2$ equals 120°. Why? Using the law of cosines, we get

$$d(P_1P_2)^2 = d(P_1P_3)^2 + d(P_2P_3)^2 + 2d(P_1P_3) \cdot d(P_2P_3) \cdot \cos 120^\circ$$

i.e., $$d(P_1P_2)^2 = (x_2 - x_1)^2 + (y_2 - y_1)^2 + 2(x_2 - x_1)(y_2 - y_1) \cdot \cos 120^\circ$$

or $$d(P_1P_2) = \sqrt{(x_2 - x_1)^2 + (y_2 - y_1)^2 + 2(x_2 - x_1)(y_2 - y_1) \cdot \cos 120^\circ}$$

***Problem 1*** Suppose that the axes intersect at an angle of θ degrees. Generalize the above argument for this case.

***Problem 2*** Show how the usual case of rectangular coordinates, i.e., perpendicular axes, is a special case of the result you found in Problem 1.

3 If a line has a slope of zero, then the line is horizontal, while a line having no slope would be a vertical line. Thus having slope zero does not mean the same thing as having no slope. The question here is one concerning the equality of zero with nothing, which is mathematically incorrect in this case.

## 2.9 Inequalities

### Questions

1 You solve $3x + 3 < x + 5$ over the set of integers and come out with $x < 1$. A student asks, "How would you check this problem as you do in an equation?" What is your response?

2 A student asks, "How do I know whether the solution of an inequality such as an absolute value inequality is the union or the intersection of the individual solution sets of each part?" How do you respond?

3 "When we write $a \leqslant b$, we mean that *a* is less than *b or a* is equal to *b*, but when we write $a \nleqslant b$, you say that this means that *a* is not less than *b and a* is not equal to *b*. I don't understand why it changes from *or* to *and*?" How do you respond to this student?

4 While studying a unit on inequalities, a student asks you if it is true that if $a > b$ and $c > d$, then $a - c > b - d$. What is your response?

5 A student asks you if it is mathematically correct to use $a \lessgtr b$ to mean *a* is less than *b* or *a* is greater than *b*. What is your response?

6 A student asks you if it is possible to use the symbolism $6 > -3 < 1$ for the idea that 6 is greater than $-3$ and $-3$ is less than 1. How do you reply?

### Answers

1 Since there is an infinite number of solutions for this problem, there is no corresponding method (substitution) for checking each solution as one would do in an equation. However, there are techniques which can be used to justify the result. For example, one might work backwards and try to show that

$$x < 1 \Rightarrow 3x + 3 < x + 5$$

as follows:

$$x < 1 \Rightarrow 2x < 2$$
$$2x < 2 \Rightarrow 3x < x + 2$$
$$3x < x + 2 \Rightarrow 3x + 3 < x + 5$$

Another technique to use is to replace the less than sign by an equality sign; so that $3x + 3 < x + 5$ is replaced by $3x + 3 = x + 5$. Solve this equation, and the result is $x = 1$. Thus in the number line, the point labeled 1 separates the line into three distinct sections. One section gives those points whose coordinates satisfy $3x + 3 < x + 5$, another section gives those points which satisfy $3x + 3 = x + 5$, and the third section gives those points whose coordinates satisfy $3x + 3 > x + 5$. By substituting an appropriate number from one of the sections, one can find the section that gives the solution set.

Notice that this is exactly the procedure one uses to solve the following problem in the plane. Find the set of ordered pairs (x, y) such that $2x + 3y < 5$.

2 For an answer to this question, refer to Section 4.1, Answer 1.

3 One way of answering this question depends on whether the student has worked with logic and truth tables. If so, then you could show the student that the negation of the statement p or q is ~p and ~q.

Another way of answering would be the following. When we write $a \nleq b$, then this means that it is not true that *a* is less than *b* or that *a* is equal to *b*; i.e., neither of them are true. Thus another way of saying this is that both $a < b$ and $a = b$ are false; i.e., $a \nless b$ and $a \neq b$. If we use the word *or*, then this leaves the option that one of them might be false but the other one is true.

If De Morgan's law has been taught to the class, then the analogy between sets and statements could be shown. That is, $\overline{A \cup B} = \overline{A} \cap \overline{B}$ corresponds to the negation of A, or B is the negation of A and the negation of B.

4 No, it is not true in general. Consider $a = 5$, $b = 2$, $c = -2$, $d = -8$. Thus $a > b$ and $c > d$, but $a - c = 7$ is not greater than $b - d = 10$.

5 This symbolism is used in some textbooks, and if defined properly, it is obviously mathematically correct. However, this is not a universally accepted conventional notation. Essentially it means that *a* and *b* are representations for two different numbers.

6 Yes, it is possible although it is not standard notation. Usually one would write $-3 < 1 < 6$. It is important for the student to understand that the symbolism $1 < 3 < 6$, as a similar example, represents *two* inequalities, namely, $1 < 3$ and $3 < 6$. Thus, $6 > -3 < 1$ represents the two inequalities $6 > -3$ and $-3 < 1$, both of which are true. Hence one would regard the inequality represented by the given symbolism as true. Thus, although the student's notation is not standard, it could certainly be used with the above agreements.

## 2.10 Logic and Proof

### Questions

1 Your class is trying to prove the following theorem. "For all real numbers b and c, $(b + c) + (-c) = b$." The proof is as follows:

| | *Statement* | *Reasons* |
|---|---|---|
| (1) | b and c are real numbers | Hypothesis |
| (2) | $b + c$ is a real number | Closure for addition axiom |
| (3) | $-c$ is a real number | Axiom of additive inverse |
| (4) | $(b + c) + (-c) = b + [c + (-c)]$ | Associative axiom of addition |
| (5) | $c + (-c) = 0$ | Axiom of additive inverses |
| (6) | $b + [c + (-c)] = b + 0$ | Substitution principle |
| (7) | $b + 0 = b$ | Addition axiom of 0 |
| (8) | $b + [c + (-c)] = b$ | Transitive property of equality |

At this time, one student raises his hand and says that he disagrees with the reason given in statement (8). He says that the reason should be the substitution principle. What do you think?

2 A ninth grade student was asked to prove the following theorem. "A positive number multiplied by a negative number equals a negative number. You may assume the following properties under multiplication:

(1) Commutative (2) Associative (3) Inverses and (4) Identity."

His proof follows.

Let x = a positive number, and let y = a positive number. I must show that $x(-y) = -(xy)$.

| | |
|---|---|
| $x(-y) = (-y) \cdot x$ | Commutative |
| $(-y) \cdot x = [-1 \cdot (y)] \cdot x$ | $(-y) = -1(y)$ |
| $[-1(y)] \cdot x = -1[y \cdot x]$ | Associative |
| $-1[y \cdot x] = -[y \cdot x]$ | $-1 \cdot z = -z$ |
| $-[y \cdot x] = -[x \cdot y]$ | Commutative |

Is the student correct in this proof?

3 A student asks you what is wrong with the following proof.

Let $c > 1$ and $c = d$.

(1) $c - 1 = d - 1$
(2) $c - 1 = -1(1 - d)$
(3) $(c - 1)^2 = (-1)^2(1 - d)^2$
(4) $(c - 1)^2 = (1 - d)^2$
(5) $(c - 1) = 1 - d$
(6) $c - 1 = 1 - c$
(7) $2c = 2$
(8) $c = 1$
(9) $1 > 1$

How do you reply?

4 A student asks, "What happens when you try to prove the $\sqrt{4}$ to be irrational as was done with $\sqrt{2}$, i.e., using 2 as a factor?" How do you respond?

## Answers

1 Either reason is correct in this case. One can show that, in many instances, the two properties are equivalent. It is easy to see that whenever the transitive property of equality can be used as a supporting reason, then the substitution principle can be used. However, the converse is not generally valid. In order to save time and effort, textbooks frequently give both properties and allow students the choice of which one

to use. For example, suppose we were given

$$x + 3 = 7 \quad \text{and} \quad x^2 + (x + 3) = 9.$$

We would usually say that, therefore, $x^2 + 7 = 9$ by the substitution principle. However, with a little work, we could show that the transitive property could be used as follows:

$$x + 3 = 7 \Rightarrow 7 = x + 3$$
$$x^2 + (x + 3) = 9 \Rightarrow x + 3 = 9 - x^2$$

Therefore, $7 = 9 - x^2$ by the transitive property or, equivalently, $x^2 + 7 = 9$.

In order to save time and effort, textbooks frequently give both properties and allow students the choice of using the one that seems more appropriate.

2 When the student writes that $(-y) = -1 \cdot (y)$ and $-1[y \cdot x] = -[y \cdot x]$, the student is really using a special case of the theorem that he is trying to prove.

3 The error is made in the fifth step; $(c - 1)^2 = (1 - d)^2$ does not imply that $c - 1 = 1 - d$. The correct way of going from (4) to (5) is:

$$(c - 1)^2 = (1 - d)^2 \Rightarrow \sqrt{(c - 1)^2} = \sqrt{(1 - d)^2}.$$
$$\sqrt{(c - 1)^2} = \sqrt{(1 - d)^2} \Rightarrow |c - 1| = |1 - d|.$$

Thus (5) should be $|c - 1| = |1 - d|$. The problem here is that $c - 1$ is positive while $1 - d$ is negative.

4 Let $\sqrt{4} = a/b$, $b \neq 0$, where $a/b$ is in lowest terms, i.e., greatest common divisor, is 1. Then $4 = a^2/b^2$, which implies that

$$4b^2 = a^2. \qquad (1)$$

Thus $a^2$ is even and therefore a is even; i.e., $a = 2 \cdot k$. Substituting in (1) we get

$$4b^2 = 4k^2 \text{ or } b^2 = k^2 \qquad (2)$$

However, (2) gives no indication of whether b is even or odd, and thus this is as far as we can go.

## 2.11 Operations on Rational and Irrational Numbers

### Questions

1 A student says that $-3$ is a whole number because it is not split up. What do you think?

2 While studying a unit on subtraction of directed numbers, a student asks you if it is possible to subtract numbers on the number line directly with vectors rather than defining subtraction in terms of addition and then doing the problem on the number line. What is your response?

3 A student asks you if there are more rational numbers than irrational numbers or are they equal in number. What is your reply?

4 A student asks, "How can you make a one-to-one correspondence between the counting numbers and the whole numbers?" What is your response?

5 How do you answer when a student in your class asks, "Since we say that $\pi = 22/7$ and $\sqrt{2} = 1.4$, why aren't these numbers rational?"

6 A student is given the problem $-7 - 2(-5)$. Her solution is: $-9(-5) = 45$. How would you explain the correct way of doing this problem to her?

7 A student is your class hands in the following work. "$12.2x^2 + 13x^2 = 25.22x^2$ because $12.2 + 13 = 25.2$ and $x^2 + x^2 = 2x^2$, so put the 2 after the 25.2, and there's the answer!" How would you react to this work?

8 A student in your class asks, "Is $\frac{100\sqrt{a}}{100} = \sqrt{a}$ an example of multiplicative inverse or multiplicative identity?" How do you respond?

9 A student says that he has trouble differentiating between the symmetric property of equality and the commutative property of addition. How could you help him?

10 For solving equalities a textbook states:

*Example* If $7x - 13 = 3x - 1$, then $7x - 13 + 13 = 3x - 1 + 13$. To use the addition property you add 13 to each member of the equality.

The student does the following:

$$7x + 13 - 13 + 13 = 3x + 13 - 1 + 13$$
$$7x + 13 = 3x + 25.$$

How would you explain to him that his method is wrong?

11 In your ninth grade algebra class, you are asked, "Does $(A + B)(C + D)$ equal $AC + BD$?" How would you answer?

12 You tell your class that 4/0 has no meaning. Some students reply that four divided by zero is still four. Since you are not dividing the four by anything, the four does not change. How do you reply?

13 A student asks, "Why doesn't $0 \div 0 = 0$, since $0 \cdot 0 = 0$?" What is your reply?

14 A student is asked to simplify $3 \div 6 + 2 - 2 \times 10$ and hands in the following:

$3 \div 6 + 2 - 2 \times 10$
$\frac{1}{2} + 2 - 2 \times 10$
$2\frac{1}{2} - 2 \times 10$
$\frac{1}{2} \times 10$
$5$

Is the student correct?

15 A student asks, "What is $0^0$?" What is your reply?

## Answers

1 For an answer to this question, read the following article by Charles Brumfiel.*

# NUMBER DEFINITIONS

By CHARLES BRUMFIEL

University of Michigan
Ann Arbor, Michigan

AFTER reading David Rappaport's article in the March 1970 MATHEMATICS TEACHER, I turned to my dictionary and found the following definitions:

*whole number* . . . an integer
*integer* . . . a whole number

Under *integer* I was referred to *complex integer*, and this definition was:

*complex integer* . . . a number $a + bi$ where $a$ and $b$ are real integers.

But real integers are not defined in my dictionary!

As badly as we handle our definitions, dictionaries do worse. This is some consolation.

I feel somewhat involved in the *whole number–counting number* terminology. University mathematicians have for centuries used *whole number* and *integer* interchangeably. I am sure that most still do so. Hence, when we began to write school texts in 1955 on the Ball State Project, I did not feel free to refer to the set of cardinal numbers, {0, 1, 2, . . .}, as the set of whole numbers. I wish now I had used the simple terminology *cardinal numbers*, but this seemed pretentious and I used *counting numbers* instead. A few years later, SMSG decreed that *whole numbers* should be used. I suggest that the phrase *counting numbers* be gracefully withdrawn from circulation.

It does seem desirable to have a short name for the positive integers, and *natural numbers* is as good as any. We should understand, however, that mathematicians have carried along two definitions for centuries. Some include zero in the set of natural numbers and some do not. This condition will persist. It is best to tell students the truth. One of the worst things you can do in teaching mathematics is to treat definitions as if they were "right" or "wrong."

I think I see the wave of the future. We shall tell children about integers in kindergarten. Having done so, we shall be able to speak of the *positive integers* and the *negative integers* from the beginning. I think this will be ill advised.

A short time ago I visited a junior high school class. A student argued that $-3$ *is* a whole number because it *is not broken*. The teacher could not satisfy him. I had the morbid desire to tell the student that $-3$ is still a whole number to most university mathematicians, excluding those who wrote for SMSG or who have been trained in the new mathematics, but I refrained. Actually, I feel that it is healthy to discuss definitions, and all teachers, including elementary teachers,

*From *The Mathematics Teacher*, April 1972, pp. 313-314. Reprinted by permission.

should utilize opportunities that arise in the classroom to do so. The task of mathematics educators is to train teachers so that they can manage such discussions.

Mr. Rappaport devoted most of his energies to fraction and rational-number terminology. I share his regret that this language is so messy. I don't know what to do about it. We should teach as many teachers as possible the standard mathematical set-theoretic definitions of ordinal numbers, cardinal numbers, and rational numbers. Every teacher who understands these definitions will realize that they are not meant for children. We must tell children something else.

One basic difficulty in choosing a school terminology for fractions and rational numbers is that so many writers wish to pretend to give precise definitions. Let's face it. The mathematical definitions are too complicated. Let's concentrate on making teachers and children comfortable without lying to them.

Why not let the terminology flow along with the use of the symbols. When the emphasis is on the use of fractions to describe the real world—and this is surely the major emphasis in the six elementary grades—just let a fraction be a symbol, written by separating two whole-number symbols by a bar and used to compare two "things" in the real world. The fraction $\frac{2}{3}$ may compare a set of boys to a set of girls, or a line segment to a unit segment, or a rectangular region to a unit region, and so on. This is the only meaning one needs to ascribe to a fraction in its use to describe the real world.

Late in the elementary school years, and in junior high school, teachers can note that for one physical situation it may be natural to use any one of many fractions to compare two given sets. The concept of *equivalent fractions* is a natural one, and we can agree that a rational number is "associated" with each equivalence class. There is no need to insist that the rational number *is* the equivalence class. This is the standard mathematical definition; but when this set-theoretic definition is formulated in a careful mathematical development, it has been presaged by a sequence of meticulous definitions, including the concepts of *ordinal number* and *ordered pair*. There is something phony about pretending to present precise definitions at this level. However, I have no objection if teachers prefer to say that the rational number *is* the equivalence class of fractions, even though the concept of an *equivalence class of symbols* is a bit hazy.

The great number-numeral controversy seems to be running its course. I shall be glad when it has faded away. Through the elementary school years, numbers are to most children simply symbols that describe the real world. This is the way we should talk about numbers. The child unblushingly adds and multiplies symbols. Don't give him a guilt feeling by telling him that he can *write* symbols but he can't *add* them, that he can *add* numbers but he can't *write* them. If you really think carefully about the process by which the game of arithmetic is abstracted from the real world, you may come to the conclusion that there is a stage when addition and multiplication really are things we do with symbols.

I do wish that someone would take his tape recorder into the heartland of America and record the definitions that children and teachers would formulate if they were asked to define whole number, natural number, integer, fraction, rational number, and so on. Mathematics educators might be greatly surprised if they knew what children and teachers really think about these things.

2 Yes, it is. The subtraction problem 3 – 5 can be represented as in the following diagram.

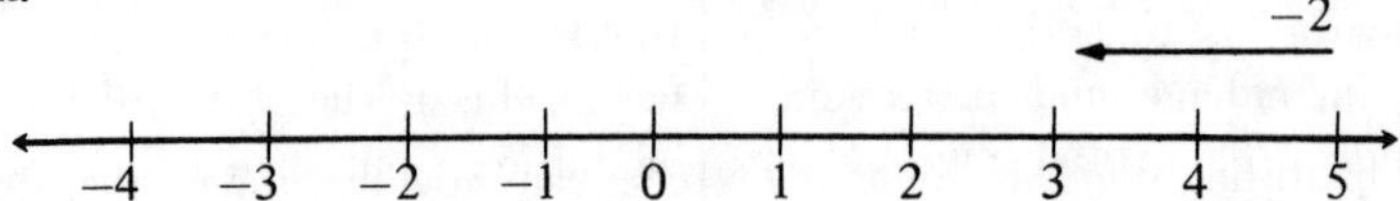


In general a – b can be represented as a vector from $b$ to $a$ as shown:

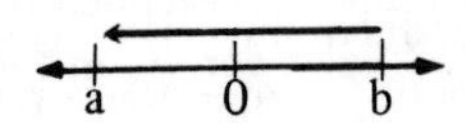


3 There are more irrational numbers than rational numbers. The rational numbers form a countable set while the real numbers form an uncountable set. If the set of irrationals formed a countable set, then the set of real numbers would be countable since the reals would then be the union of two countable sets, which is again countable. (Thus the set of irrationals is not countable.)

An intuitive way of showing this to students would be as follows. Let r be any rational number and i be any irrational number. Then r + i will always be an irrational number. Let us fix r and let i run through the set of all irrational numbers. If we repeat this process for each rational number, then we can say that for each rational number, there are an infinite number of irrational numbers. Since there are an infinite number of rational numbers, we would get an infinite number of infinite sets where each set consists of irrational numbers.

4 Consider $n \to n + 1$. This is a one-to-one correspondence between the counting numbers and the whole numbers.

5 If we say that $\pi = 22/7$ and $\sqrt{2} = 1.4$, then we have made incorrect statements. What we should say is that 22/7 is a rational approximation of $\pi$ and that 1.4 is a rational approximation of $\sqrt{2}$. In symbols this is usually denoted by $\pi \approx 22/7$ and $\sqrt{2} \approx 1.4$. $\pi$ and $\sqrt{2}$ are irrational numbers, but the value of these two numbers can be approximated by rational numbers. This is what is meant by $\pi = 22/7$ and $\sqrt{2} = 1.4$.

6 In problems of this nature, the convention is to do the operations in this order:

(1) Numbers in parentheses first, etc.

(2) Multiplication and division from left to right in that order, etc.

(3) Addition and subtraction from left to right in that order, etc.

Thus, $-7 - 2(5) = -7 - (-10) = -7 + 10 = 3$.

7 Of course, the student is incorrect. One might show the student how the distributive property of multiplication over addition is used in this situation:

$$12.2x^2 + 13x^2 = (12.2 + 13)x^2 = (25.2)x^2 = 25.2x^2.$$

Another approach might be to ask the student to do a simple problem such as

$2x + 2x$. Two groups of x plus two groups of x are four groups of x. Thus $2x + 2x = 4x$. Using the student's method:

$$2 + 2 = 4 \quad \text{and} \quad x + x = 2x, \qquad \text{so} \quad 2x + 2x = 8x.$$

By substituting values in for x, the teacher can show the student that this answer is incorrect.

8 The answer to this question involves more than one property. Frequently the problem would be stated something like the following. "Solve for $\sqrt{a}$ in the problem below."

$100\sqrt{a} = y.$

| | |
|---|---|
| $100\sqrt{a} = y$ | Given |
| $\frac{1}{100}(100\sqrt{a}) = \frac{1}{100} \cdot y$ | Multiplication property of equality; i.e., if a, b, c, d are real numbers such that a = b and c = d, then ac = bd. |
| $\left(\frac{1}{100} \cdot 100\right)\sqrt{a} = \frac{1}{100}y$ | Associative property of multiplication |
| $1 \cdot \sqrt{a} = \frac{1}{100}y$ | Multiplicative inverse |
| $\sqrt{a} = \frac{1}{100}y$ | Multiplicative identity |

9 One way of helping this student would be as follows. The commutative property of addition says that the order in which we add two numbers makes no difference since the result will be the same number. For example, $3 + 4 = 4 + 3 = 7$. Using the same numbers, the symmetric property of equality says that if $3 + 4 = 4 + 3$, then $4 + 3 = 3 + 4$. Notice that the commutative property involves switching the order of the addends while the symmetric property involves switching the whole expression.

10 You might rephrase the expression from the book so that it reads, "To use the addition property, you add 13 to each side of the equality." The student is interpreting the word *member* to mean each monomial in the equation.

11 No, $(A + B)(C + D) \neq AC + BD$. If one uses the distributive property twice, one can see that

$$(A + B)(C + D) = (A + B)C + (A + B)D = AC + BC + AD + BD.$$

Another way of showing this is to ask the student to substitute numbers in for A, B, C, D. For example, if $A = 1$, $B = 2$, $C = 3$ and $D = 4$,

$$(A + B)(C + D) = 3 \cdot 7 \quad \text{or} \quad 21,$$
$$AC + BD = 3 + 8 = 11$$

Since $11 \neq 21$, then $(A + B)(C + D) \neq AC + BD$.

12 If $\frac{4}{0} = 4$, then it should be the case that $4 = 0 \cdot 4$. Since $4 \neq 0$, then it cannot be true that $\frac{4}{0} = 4$. Since $\frac{4}{0}$ is equivalent to $4 \div 0$ and division can also be interpreted as repeated subtraction, then another way to look at this problem is as follows. How many times can we subtract zero objects from 4 objects until the 4 objects are gone. One could never accomplish this task. Thus $\frac{4}{0}$ does not represent any number.

13 It is true that $0 \cdot 0 = 0$, but the logic used to infer $0 \div 0 = 0$ could also be used to infer $0 \div 0 = r$ since $0 \cdot r = 0$. Thus, how does one decide which number r should be used. It is this lack of uniqueness that has led mathematicians to say that $0 \div 0$ is undefined.

14 No, the student is not correct. According to convention, multiplication and division are done first from left to right in whatever order they occur. Then addition and subtraction are done in the same manner. Thus, according to this convention, $3 \div 6 + 2 - 2 \times 10$ should be done as follows.

$$3 \div 6 + 2 - 2 \times 10 = \tfrac{1}{2} + 2 - 20 = 2\tfrac{1}{2} - 20 = -17\tfrac{1}{2}.$$

15 Probably at this level the best answer would be to say that $0^0$ is an undefined expression. However, the case is not as clear cut as it might seem, and the two articles cited below give background information on this expression.

Herbert Vaughan, "The Expression '$0^0$'," *The Mathematics Teacher*, Vol. 63 (1970), pp. 111-2.

Also consult Answer 5 in Section 4.4

## 2.12 Number Theory and Average

### Questions

1 You make the statement that there is no last prime. A student says there is. The student reasons that since the distance between the primes increases as the numbers get larger, then it follows that at some point the space becomes infinitely long, which is the same as not having an end. How would you handle this situation?

2 A student asks you why we only consider positive numbers as primes and do not consider negative numbers. What is your reply?

3 In a discussion on rates a student says, "If I walk 4 mph for one mile, I'll have to run 8 mph the second mile in order to average 6 mph for 2 miles. Right?" How do you reply?

4 "If I know all the prime numbers which precede a certain given prime, is there any way to generate additional primes? Is there any way, other than trial and error, to find the prime number which follows a given prime?" How do you respond?

## Answers

1 This student is making an unwarranted conclusion. The fact that the distance between numbers "gets longer" does not mean that at some point this distance becomes infinitely long. For example, the distance between the numbers 0, 1, 3, 6, 10, 15, 21, 28, ... gets longers as one considers consecutive pairs "further out" in the sequence. However, at no stage does this distance become infinite. The teacher might suggest that the student consider (intuitively) that distance *between* numbers cannot possibly be infinite because, in order to be *between*, the distance must be finite.

2 The study of prime numbers began with the Greeks long before negative integers were invented. Thus, historically, primes were developed with positive integers only, and it has been customary to follow, at least at the elementary level, this historic precedent.

However, in advanced mathematics, such as abstract or modern algebra, it is the case that a more general approach is taken. In this case, one deals with irreducibles rather than primes, and one studies such notions as unique factorization domains, principal ideal domains, Euclidian domains, algebraic and Gaussian integers, norms, etc. For more information on this approach consult:

Fraleigh, *A First Course in Abstract Algebra*, Addison-Wesley Publishing Co., Reading, Ma., 1969, Section 32, 33 and 34.

3 No, the student is not correct. Average speed is defined as follows.

$$\text{Average speed} = \frac{\text{total distance}}{\text{total time}} \tag{1}$$

Thus substituting your values in (1), we get

$$\text{Average speed} = \frac{2}{(1/4) + (1/8)} = \frac{2}{3/8} = \frac{16}{3} = 5\frac{1}{3}.$$

In order to average 6 mph for the two miles, you would have to solve for x as follows. Let x denote the speed for the second mile. Then

$$6 = \frac{2}{(1/4) + (1/x)}$$

$$6 = \frac{8x}{4 + x}$$

$$8x = 24 + 6x$$

$$x = 12.$$

Thus you must run at 12 mph for the second mile.

4 If $n_1, n_2, ..., n_k$, is a list of primes, then the integer

$$n_1 n_2 \cdots n_k + 1$$

is not divisible by any one of the primes $n_1, n_2, ..., n_k$. Thus it is either a prime that is not in the given list, a power of a prime that is not in the given list, or a composite integer divisible by a prime that is not in the given list. For example, given the list of primes 3, 5, 7, then $3 \cdot 5 \cdot 7 + 1 = 106 = 2 \cdot 53$ leads to the new prime 53.

One could also use trial and error with a computer. In order to find out if x is a prime number, one has only to divide by prime numbers less than $\sqrt{x}$, which simplifies the work considerably but still leaves a great deal of computation if x happens to be a large number.

For more information on primes consult:

(1) Stanley Ogilvy, *Tomorrows's Math*, 2nd ed., Oxford University Press, New York, 1972, Chap. 5.

(2) Hans Rademacher and Otto Toeplitz, *The Enjoyment of Mathematics*, Princeton University Press, Princeton, N.J., 1957, Chap. 1.

## 2.13 Operations, Algorithms, Properties of Operations

### Questions

1 A student asks you if addition is distributive over multiplication. What do you say?

2 A student asks you why $(-1)(-1) = 1$. The student says that it doesn't make sense. How would you help this student?

3 A student asks you if $\frac{1}{a+b} = \frac{1}{a} + \frac{1}{b}$. How do you reply?

4 "What is the difference between the additive inverse of a number and the negative of a number?" What do you tell this student?

5 A student hands in the following work.

$$(6t)\sqrt{3} = 6\sqrt{3} \cdot t\sqrt{3}$$

How would you help this student?

6 In discussing combination of like terms, the teacher has used the analogy of adding numbers with like units (i.e., 5 days + 2 days). Students have just been told that they can't add 5 days and 2 hours without changing one of the units. When asked what to do with $2x + 5y$, one student wants to know how he can change x and y so that they can be added. What is your response?

7 A student says that $\frac{1}{2}(18\sqrt{3}) = \frac{1}{2}(18) \cdot \frac{1}{2}\sqrt{3}$. How would you explain to your student that he is incorrect?

8 Students in your class are asked to simplify $(3 + T) - (T - 5)$. A student replies:

$$\begin{aligned}(3 + T) - (T - 5) &= 3 + (T - T) - 5 \\ &= 3 - 5 \\ &= -2.\end{aligned}$$

Is the student correct?

## Answers

1 If addition was distributive over multiplication, then

$$a + (b \cdot c) = (a + b) \cdot (a + c) \qquad (1)$$

However, (1) is not true for all real numbers a, b, c. For example, if $a = 1, b = 2, c = 3$, then (1) becomes

$$1 + (2 \cdot 3) = (1 + 2)(1 + 3),$$

which simplifies to the *false* statement $7 = 12$. However, we might be tempted to try to find out under what conditions (1) would be true.

$$\begin{aligned}a + (b \cdot c) = (a + b)(a + c) &\Rightarrow a + b \cdot c \\ &= a^2 + ab + ac + bc \Rightarrow a = a^2 + ab + ac.\end{aligned}$$

If $a \neq 0$, then $1 = a + b + c$. Thus (1) is true when either $a = 0$ and b and c are any real numbers, or when $a \neq 0$ and $a + b + c = 1$.

Note that it might be pointed out to the student that even though (1) is not generally valid in our algebraic system, it is valid in a mathematical system called Boolean Algebra. Perhaps at this point some students might want to explore this interesting mathematical system.

2 For an answer to this question, read:

John C. Peterson, "Fourteen Different Strategies for Multiplication of Integers or Why $(-1)(-1) = +1$," *The Arithmetic Teacher*, May 1972, pp. 396-403.

3 The correct answer is "No." To show this, ask the student to substitute $a = b = 1$, which should lead the student to the *false* statement $\frac{1}{2} = 2$.

Actually, for *any* choice of *a* and *b* for which $\frac{1}{a+b}$ and $\frac{1}{a} + \frac{1}{b}$ are defined, the resulting expressions will be different. To show this might require more sophistication than can be expected at this level, but one method is the following.

We seek numbers *a* and *b* such that $\frac{1}{a+b} = \frac{1}{a} + \frac{1}{b}$. Now by definition,

$$\frac{1}{a} + \frac{1}{b} = \frac{a+b}{ab},$$

Hence one is looking for numbers *a* and *b* such that the above equation is true. Or equivalently,

$$(a + b)^2 = ab \quad \text{or} \quad a^2 + 2ab + b^2 = ab, \quad \text{or}$$
$$a^2 + ab + b^2 = 0. \qquad (1)$$

Now, pick *any* number b, b $\neq$ 0, and substitute into (1). Then one obtains a quadratic in *a*, which gives

$$a = \frac{-b \pm \sqrt{b^2 - 4b^2}}{2} \quad \text{or} \quad a = \frac{-b \pm \sqrt{-3b^2}}{2}.$$

It is clear now that there is *no* corresponding (real) number *a*.

4 For an answer to this question, read the following article by Gary Musser.*

# ADDITIVE INVERSES and NEGATIVES— TWO DIFFERENT CONCEPTS

By GARY L. MUSSER

Northern Illinois University
De Kalb, Illinois

CERTAINLY one of the best-known statements in arithmetic is the statement "The product of two negatives is positive." This expression is generally associated with the study of the ring of integers, but it also applies in the fields of rational and real numbers. Usually a ring-theoretic proof of the fact $(-a)(-b) = ab$ is provided to prove the statement quoted above; however, the equation $[-(-5)](-3) = (-5)3$ is an instance of $(-a)(-b) = ab$, but certainly the equation $[-(-5)](-3) = (-5)3$ is not described by the sentence "The product of two negatives is positive." Sometimes $a$ and $b$ are restricted to be positive integers; thus $(-a)$ and $(-b)$ are negative integers and all is well. Also, $(-x)$ is sometimes referred to as "the negative of $x$," and then, in this context, $(-a)(-b) = ab$ is considered to be an instance of "The product of two negatives is positive." However, it seems that more needs to be said. It is the purpose of this article to clarify the distinction between the two statements "The product of two negatives is positive" and "The product of the additive inverses of $a$ and $b$ is $ab$." I begin by establishing the fundamental notions that form the foundation of my discussion.

## Rings and Additive Inverses

A set $R$ together with two binary operations addition and multiplication defined on $R$ is called a *ring* if the following familiar properties are satisfied.

R1: $(a + b) + c = a + (b + c)$ for all $a, b, c \in R$.

R2: There exists an element $0 \in R$ such that $a + 0 = a$ for all $a \in R$.

R3: For each $a \in R$ there exists an element $(-a) \in R$ such that $a + (-a) = 0$.

R4: $a + b = b + a$ for all $a, b \in R$.

R5: $(ab)c = a(bc)$ for all $a, b, c \in R$.

R6: $a(b + c) = ab + ac$ and $(b + c)a = ba + ca$ for all $a, b, c \in R$.

The integers, rational numbers, real numbers, and complex numbers, together with their usual additions and multiplications, are rings.

Using the properties of a ring and the (provable) fact that $a \cdot 0 = 0 \cdot a = 0$, the following detailed ring-theoretic proof of the fact $(-a)(-b) = ab$ can be given.

*From *The Mathematics Teacher*, April 1972, pp. 301-304. Reprinted by permission.

THEOREM. $(-a)(-b) = ab$ *for all* $a, b \in R$.

| *Proof.* $(-a)(-b)$ | Reasons: |
|---|---|
| $= (-a)(-b) + 0$ | R2 |
| $= (-a)(-b) + a \cdot 0$ | $a \cdot 0 = 0$ for all $a$ |
| $= (-a)(-b) + a[(-b) + b]$ | R3 and R4 |
| $= (-a)(-b) + [a(-b) + ab]$ | R6 |
| $= [(-a)(-b) + a(-b)] + ab$ | R1 |
| $= [(-a) + a](-b) + ab$ | R6 |
| $= 0 \cdot (-b) + ab$ | R4 and R3 |
| $= 0 + ab$ | $0 \cdot a = 0$ for all $a$ |
| $= ab$ | R4 and R2 |

In words, this theorem says that "the product of the additive inverses of $a$ and $b$ is $ab$." Applied to the ring of integers, it shows that $(-5)(-3) = 5 \cdot 3$ where $-5$ and $-3$ are viewed as additive inverses of 5 and 3 respectively. To gain more insight into the statement "The product of two negatives is positive," the concepts of a partial order and a partially ordered ring will be considered next.

### Partially Ordered Rings

A binary relation "$\leq$" on a set $R$ is called a *partial order* if it satisfies the following properties where $a, b, c \in R$:

P1: $a \leq a$ for all $a \in R$.
P2: If $a \leq b$ and $b \leq a$, then $a = b$.
P3: If $a \leq b$ and $b \leq c$, then $a \leq c$.

In the ring of integers, the usual order, $a \leq b$ if $b - a \in \{0, 1, 2, \cdots\}$, is an example of a partial order. Graphically, $a \leq b$ holds if $a$ coincides with $b$ or $a$ is to the left of $b$ on the usual integer number line. Also, in any collection of sets, the relation "is a subset of" satisfies P1, P2, and P3; hence it is a partial order.

A ring $R$ together with a partial order on $R$ is called a *partially ordered ring* if for all $a, b, c \in R$ the following computational properties are satisfied:

PR1: If $a \leq b$, then $a + c \leq b + c$.
PR2: If $a \leq b$ and $0 \leq c$, then $ac \leq bc$ and $ca \leq cb$.

Notice that properties PR1 and PR2 relate the ring operations with the order relation just as distributivity connects addition with mutliplication in R6. The integers, the rational numbers, and the real numbers with their usual orders are partially ordered rings. The elements $a$ such that $0 \leq a$ are called *positive* elements and those satisfying $a \leq 0$ are called *negative* elements. (Note: I follow Fuchs [1963] in considering zero to be both positive and negative. Nonzero positive elements are called strictly positive, denoted by $0 < a$, and nonzero negative elements, denoted by $a < 0$, are called strictly negative.)

The integers, rational numbers, and real numbers as partially ordered rings satisfy an additional property called the *trichotomy property:*

P4: For all $a \in R$, one and only one of the following holds—

$$a < 0 \text{ or } a = 0 \text{ or } 0 < a.$$

A partially ordered ring $R$ that also satisfies the trichotomy property is called a *linearly ordered ring*, since the order among the elements of $R$ can be displayed by the points of a line. Not all partially ordered rings are linearly ordered rings; in other words, there are partially ordered rings that contain nonzero elements that are neither positive nor negative. Later, using examples, an appeal to this fact will be made to clarify the distinction between the terms *negative* and *additive inverse*.

### Negatives and Additive Inverses

Let $R$ be a partially ordered ring. If $a$ is a negative element, then $a \leq 0$. By PR1, $a \leq 0$ implies that $a + (-a) \leq 0 + (-a)$, or $0 \leq (-a)$. Thus, if $a$ is negative, its additive inverse is positive. Now let $a$ and $b$ be negative; hence, $(-a)$ and $(-b)$ are positive. Since $R$ is a ring, we have $(-a)(-b) = ab$. Furthermore, it follows from PR2 that the product of two positives is positive; hence, $(-a)(-b)$ is positive. Therefore the product $ab$ of two negative elements $a$ and $b$ is positive. It is apparent from this proof that in a partially ordered ring the fact that the product of two negatives is positive is a *consequence* of the ring-theoretic fact $(-a)(-b) = ab$. Now let us consider examples that illustrate how the notions of additive inverses and negatives differ.

Our first example, though nonstandard, clearly illustrates the necessity for distinguishing between additive inverses and negatives. Define a relation "$\leq\leq$" on the ring of integers by having $a \leq\leq b$ if $b - a \in \{0, 2, 4, \cdots\}$. Then it is easy to show that "$\leq\leq$" is a partial order for the integers; moreover, the integers form a partially ordered ring with respect to "$\leq\leq$." As usual, we say that $a$ is 'positive' if $0 \leq\leq a$ and that $a$ is 'negative' if $a \leq\leq 0$. Therefore, $\{0, 2, 4, \ldots\}$ is the set of all 'positive' integers, and $\{0, -2, -4, \ldots\}$ is the set of all 'negative' integers (see fig. 1). It still follows from the ring-theoretic proof that $(-5)(-3) = 15$; however, $(-5)(-3) = 15$ is *not* an instance of the statement "The product of two 'negatives' is 'positive,'" since neither $(-5)$ nor $(-3)$ is negative (nor is 15 positive) with respect to the partial order "$\leq\leq$."

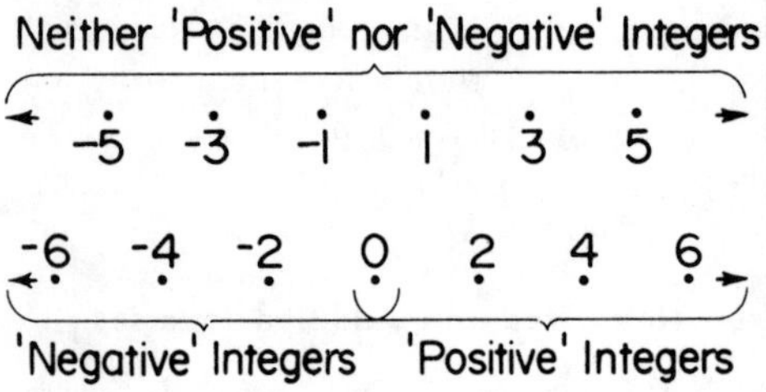

Neither 'Positive' nor 'Negative' Integers

-5 -3 -1 1 3 5

-6 -4 -2 0 2 4 6

'Negative' Integers 'Positive' Integers

Fig. 1

For a second example, let $C(R)$ denote the set of all continuous functions having the field of real numbers as their domain and range. For $f, g \in C(R)$, if $f + g$ and $f \cdot g$ are defined by $(f + g)(x) = f(x) + g(x)$ and $(f \cdot g)(x) = f(x)g(x)$ respectively, then $C(R)$ together with this addition and multiplication is a ring. Furthermore, if an order on $C(R)$ is defined by $f \leq g$ whenever $f(x) \leq g(x)$ for all real numbers $x$, then the ring $C(R)$ together with this order relation is a partially ordered ring. Graphically, $f \leq g$ holds if and only if the graph of $f$ is never above the graph of $g$. Hence, a function $f$ is positive if $0 \leq f(x)$ for all real numbers $x$ and negative if $f(x) \leq 0$ for all real numbers $x$. Clearly, there are many functions in $C(R)$ that are neither positive nor negative; for example, $f(x) = x$ is such a function. In fact, any function whose graph has points both above and below the $x$-axis is such a function (fig. 2). Now consider

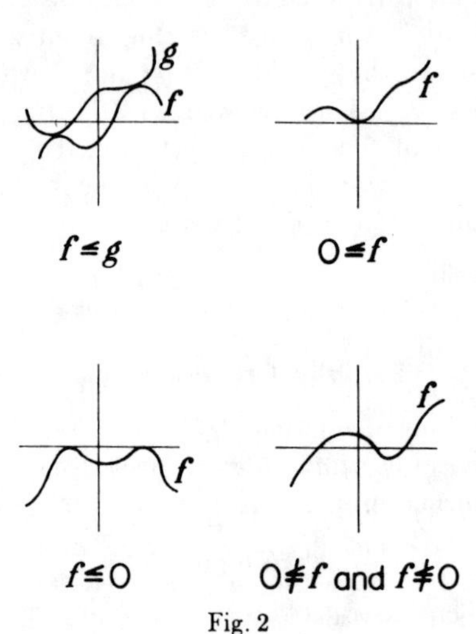


Fig. 2

the functions $f$ and $g$ defined by the equations $f(x) = x$ and $g(x) = 5 - x$. Then $(-f)$ and $(-g)$ are defined by $(-f)(x) = -x$ and $(-g)(x) = x - 5$, and $f \cdot g$ is given by $(f \cdot g)(x) = x(5 - x) = 5x - x^2$. Now $(-f)(-g) = f \cdot g$; however, it is easily seen that neither $(-f)$ nor $(-g)$ is negative, nor is $f \cdot g$ positive (fig. 3).

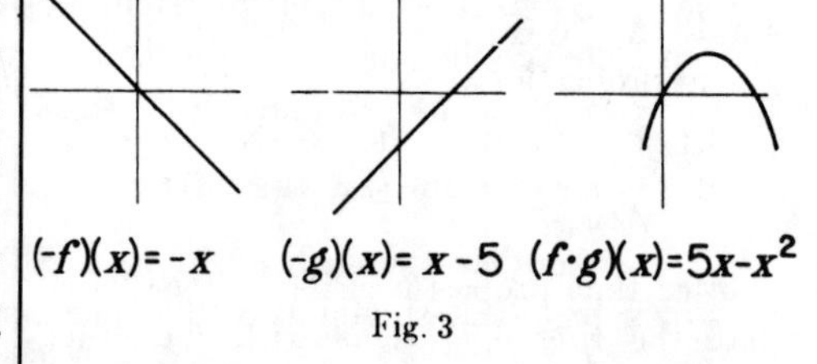


Fig. 3

Here again, the statement "The product of two negatives is positive" is not an appropriate description of the equation $(-f) \cdot (-g) = f \cdot g$.

There are many other examples of partially ordered rings which clearly distinguish between negatives and additive inverses. For example, let $R$ be the ring of 2-by-2 matrices over the real numbers and define "$\leq$" by

$$\begin{bmatrix} a & b \\ c & d \end{bmatrix} \leq \begin{bmatrix} e & f \\ g & h \end{bmatrix}$$

if $a \leq e$, $b \leq f$, $c \leq g$, and $d \leq h$.

Also, let $R$ be the field of complex numbers, and define "$\leq$" by $a + bi \leq c + di$ if $a \leq c$ and $b \leq d$. Both these examples are partially ordered rings, and as in the preceding examples, they are not linearly ordered rings; so they both contain elements that are neither positive nor negative.

The above discussion indicates the importance of distinguishing between negatives and additive inverses, and, furthermore, it demonstrates the need to use precise language when discussing these two concepts.

## Epilogue

To illustrate the necessity for differentiating between statements involving additive inverses and negatives, partially ordered rings have been used. Appealing to partially ordered rings is quite natural because they play such a fundamental role in mathematics. However, the critical reader may have observed that it was necessary to appeal only to an algebraic structure that is "between" a ring and a partially ordered ring.

To elaborate, let $R$ be a ring under addition and multiplication, and let $P$ be a nonempty subset of $R$ that is closed under multiplication. Then call the elements in $P$ positive and their additive inverses negative. Now, as before, we can prove that the product of two negatives is positive. This reasoning is the kind that is sometimes used to justify the statement "The product of two negatives is positive" in the ring of integers. There, the natural numbers are called positive integers and their additive inverses are called negative integers. Although this is a correct approach, a partially ordered ring structure seems to be the appropriate mode of justification because of its importance among the structures of mathematics.

### REFERENCES

Birkhoff, G., and S. MacLane. *A Survey of Modern Algebra*. 3d ed. New York: Macmillan Co., 1965.

Fuchs, L. *Partially Ordered Algebraic Systems*. New York: Pergamon Press, 1963.

Halmos, P. *Naive Set Theory*. Princeton: Van Nostrand Co., 1960.

5 The student is assuming that multiplication distributes over multiplication, which is not valid in our algebraic system; i.e., $a \cdot (b \cdot c) = (a \cdot b)(a \cdot c)$. You might try to convince the student of this fact by asking the student if $2 \cdot (3 \cdot 4) \neq (2 \cdot 3) \cdot (2 \cdot 4)$. Since $24 \neq 48$, then

$$a \cdot (b \cdot c) \neq (a \cdot b)(a \cdot c).$$

Another approach might be to show that if $(6t)\sqrt{3} = 6\sqrt{3} \cdot t\sqrt{3}$, then the right hand side should name the same numbers as the left hand side for all values of t. If $t = 1$, then $(6t)\sqrt{3} = 6\sqrt{3}$ while $6\sqrt{3} \cdot t\sqrt{3} = 6\sqrt{3}\sqrt{3} = 18$. In general,

$$6\sqrt{3} \cdot t\sqrt{3} = (6t)(\sqrt{3} \cdot \sqrt{3}) = 6t \cdot 3 = 18t$$

and $18t \neq 6t\sqrt{3}$.

6 For adding 5 days and 2 hours, we know that there is a definite relationship between days and hours; i.e., 5 days = 120 hours. Thus 5 days and 2 hours = 120 hours and 2 hours = 122 hours. However, if we are just given x and y, then without additional information we do not know of the relationship between these two quantities. Thus the only thing we can do is to leave them as they are. For more information on this subject, consult:

Hassler Whitney, "The Mathematics of Physical Quantities," *The Mathematical Monthly*, Vol. 75, No. 2, Part I. (Part II appears in the next issue.)

7 One way of explaining to the student that he is incorrect is to write the problem as follows.

$$\frac{1}{2}(18\sqrt{3}) = \frac{1}{2} \cdot (18 \cdot \sqrt{3}).$$

Since the numbers involved are real numbers, and since multiplication is associative,

$$\frac{1}{2} \cdot (18 \cdot \sqrt{3}) = \left(\frac{1}{2} \cdot 18\right)\sqrt{3} = 9\sqrt{3}.$$

However, by the student's method $\frac{1}{2}(18\sqrt{3}) = (9/2)\sqrt{3}$.

Another way of convincing the student that he is wrong is to write $\frac{1}{2}(18\sqrt{3})$ as $\frac{18\sqrt{3}}{2}$ or $(18\sqrt{3}) \div 2$. Then you might explain to the student that there are 18 square roots of 3 and we want to divide them by 2. Thus, we should get 9 square roots of 3 for the answer.

The basic procedure here is to show the student that multiplication is not distributive over multiplication. Thus, you might ask the student if

$$2 \cdot (3 \cdot 4) = (2 \cdot 3)(2 \cdot 4).$$

Hopefully, the student would see that since $24 \neq 48$, then

$$a \cdot (b \cdot c) \neq (a \cdot b) \cdot (a \cdot c).$$

8 No, the student is not correct.

$$\begin{aligned}(3 + T) - (T - 5) &= (3 + T) + (-(T - 5))\\ &= (3 + T) + (5 - T)\\ &= (3 + 5) + (T - T)\\ &= 8\end{aligned}$$

The student is trying to use a form of the associative property in a problem involving subtraction, but subtraction is not associative.

## 2.14 Polynomials and Rational Expressions

### Questions

1 While dividing $15x^4 + 9x^3 + 4x^2 + 1$ by $x^2 - 1$, one student writes that the quotient is $15x^2 + 9x + 19$ and the remainder is $\frac{9x + 20}{x^2 - 1}$. Is the student correct?

2 A student makes the following statement. "You said that $2x^2 + 2x$ is a polynomial, but $2(x^2 + x)$ is not a polynomial even though it is equivalent to $2x^2 + 2x$. I don't understand?" How could you help this student?

3 You make the statement that the polynomial $x^2 - 2$ cannot be factored. One student says, "Yes, it can because $x^2 - 2 = (x - \sqrt{2})(x + \sqrt{2})$." What reply would you make?

4 The following item is on an algebra test:

Which of the following are not rational expressions and why.

(1) $x$ (2) $\frac{1}{x}$ (3) $\frac{1}{x^2}$ (4) $\frac{1}{|x|}$ (5) $\frac{1}{x^2 + 2}$

A student in your class circles number (4) $\frac{1}{|x|}$ and says that it is not a rational expression because it has two values. How would you handle this situation?

5 A student in your class asks, "Do rational operations performed with irrational numbers always result in irrational numbers?" How would you answer?

### Answers

1 No, the student is not correct. The division algorithm for polynomials states that if A and B are polynomials where B is not the zero polynomial, then there exist unique polynomials, Q and R, such that $A = B \cdot Q + R$, where R is either the zero polynomial or where the degree of R is less than the degree of B. If the above is multiplied by $1/B$, then one obtains the quotient

$$\frac{A}{B} = Q + \frac{R}{B}$$

or, in our problem,

$$\frac{15y^4 + 9y^3 + 4y^2 + 1}{y^2 - 1} = 15y^2 + 9y + 19 + \frac{9y + 20}{y^2 - 1}.$$

2 A polynomial over the set of integers is defined to be a monomial or the algebraic sum of two or more monomials over the set of integers. A monomial over the integers is defined to be either a product of integers or a product of integers and variables with the exponents of the variables being positive and integral. The expression $2(x^2 + x)$ is thus not a polynomial by definition. However, $2(x^2 + x)$ is a

polynomial expression where the term polynomial expression is defined to be any polynomial or any expression that can be formed by using two or more polynomials and indicating the operations by addition or multiplication. Thus all polynomials are polynomial expressions, but not all polynomial expressions are polynomials.

3 The problem here is that the teacher made a statement leaving out a necessary restriction. The teacher meant to say that $x^2 - 2$ cannot be factored over the set of rational numbers. The student is correct since $x^2 - 2$ can be factored over the set of real numbers.

4 The student is correct in saying that $1/|x|$ is not a rational expression, but the reason given by the student is incorrect. $1/|x|$ is not a rational expression because $|x|$ is not a polynomial. One should question the student further as to what he means by *two values.*

5 No, not always. Consider the following expression $\sqrt{3} - \sqrt{3}$ which equals the rational number 0. Another example would be $\sqrt{2} \cdot \sqrt{2} = 4$.

## 2.15 Radicals and Square Roots

### Questions

1 While discussing cube roots, you give your class the following problem.

$$\sqrt[3]{16} + \sqrt[3]{-16} = ?$$

A student says that the answer is zero because $16 - 16 = 0$. Is the student right?

2 A student asks, "Is it correct to write $2^{1/16}$ as

$$\sqrt{\sqrt{\sqrt{\sqrt{2}}}}$$

What is your reply?

3 While showing your students how to find a square root using the standard algorithm, one student asks you why it works. What do you tell the student?

4 In response to the question, "How do you define $\sqrt{x^2}$?" a student says that $\sqrt{x^2} = x$. Is this student correct?

5 A student asks, "Why isn't $\sqrt{3} + \sqrt{12} = \sqrt{15}$?" How do you reply?

6 You have worked the following problem on the board. "Simplify $\sqrt{11/3x^2}$ and you get your final answer of $\sqrt{33}/3x$." However, one of your students says that the final answer should be $\sqrt{33}/3|x|$. What is your reply?

7 While studying an algorithm for finding square roots, a student asks you if there is an algorithm for finding the cube of a number. What is your response?

## Answers

1 The student's answer is correct, but the method given is not. Consider the following problem:

$$\sqrt{16} + \sqrt{-16} = ?$$

According to the student's method, the answer should be zero because $16 - 16 = 0$. However, $\sqrt{16} = 4$ while $\sqrt{-16} = 4i$ and $4 + 4i \neq 0$.

The student will get the correct answer in the case where the root is the same odd number. This is reflected in the example

$$\sqrt[n]{k} + \sqrt[n]{-k} = 0 \qquad \text{when } n = 2t + 1,\ t \in I \text{ and } k \in I.$$

2 Yes, it is correct as illustrated below.

$$2^{1/16} = (2^{1/8})^{1/2} = \sqrt{2^{1/8}}$$

$$2^{1/8} = (2^{1/4})^{1/2} = \sqrt{2^{1/4}}$$

Thus $2^{1/16} = \sqrt{\sqrt{2^{1/4}}}$,

$$2^{1/4} = (2^{1/2})^{1/2} = \sqrt{\sqrt{2}}.$$

Thus $2^{1/16} = \sqrt{\sqrt{\sqrt{2}}}$.

3 For an answer to this question, refer to Section 1.13, Answer 1.

4 No, this student is not correct. $\sqrt{x^2}$ means that nonnegative number t such that $x^2 = t^2$. If $x = -5$, then according to the student's definition $\sqrt{(-5)^2} = -5$, but $-5$ is a negative number. The correct definition should be $\sqrt{x^2} = |x|$.

5 What you are really saying is that $\sqrt{a} + \sqrt{b} = \sqrt{a + b}$ for all a, b. However, this is not correct. Consider $a = 4$ and $b = 4$, then $\sqrt{a} + \sqrt{b} = 4$ while $\sqrt{a + b} = \sqrt{8} = 2\sqrt{2}$, which is less than 4. Thus, in general, $\sqrt{a} + \sqrt{b} \neq \sqrt{a + b}$.

***Problem*** Under what conditions will the student's method work?

6 Simplifying the expression $\sqrt{11/3x^2}$ is usually done as illustrated:

$$\sqrt{\frac{11}{3x^2}} = \frac{\sqrt{11}}{\sqrt{3x^2}} \tag{1}$$

$$\frac{\sqrt{11}}{\sqrt{3x^2}} = \frac{\sqrt{11}}{\sqrt{3}\sqrt{x^2}} \tag{2}$$

$$\frac{\sqrt{11}}{\sqrt{3}\sqrt{x^2}} = \frac{\sqrt{33}}{3\sqrt{x^2}} \tag{3}$$

$$\frac{\sqrt{33}}{3\sqrt{x^2}} = \frac{\sqrt{33}}{3|x|} \tag{4}$$

Thus, the student is correct. However, after a set of exercises using the definition of absolute value are assigned to students, textbooks frequently specify that henceforth we shall assume that all variables are positive unless otherwise specified. If this is the situation with the given problem, then the teacher is justified in writing $\sqrt{33}/3x$ in (4) above. However, this restriction should be pointed out to the students in order to avoid confusion.

7 One algorithm for finding cube roots is given below. This algorithm is similar to one of the algorithms for finding square roots and is based on the binomial theorem; i.e.,

$$(a + b)^3 - a^3 = (3a^2 + 3ab + b^2) \cdot b.$$

Thus, to find the cube root of 9685, one proceeds as follows.

| | 9 685.000 (21.3) |
|---:|:---|
| | 8 |
| 1200 | 1 685 |
| 60 | |
| 1 | 1 261 |
| 1261 | |
| 132300 | 424 000 |
| 1890 | |
| 0 | 402 597 |
| 134199 | 21 403 |
| | etc. |

One might tell the student that there are other methods for finding cube and higher order roots such as using logarithms, Newton's method and Horner's method which the student will learn in more advanced courses in mathematics. For example, using logarithms:

$$\log \sqrt[3]{9.685} = \frac{1}{3} \log 9.685$$
$$= \frac{1}{3} \cdot (3.9861)$$
$$\log \sqrt[3]{9.685} = 1.3287$$
$$\sqrt[3]{9.685} = 21.3$$

## 2.16 Sets

### Questions

1 In your ninth grade class, you make the statement that the empty set is a subset of every set. A student doesn't believe it. How could you help this student?

2 A student says that a set is finite if it has a largest element. Do you agree?

3 A student asks, "Are any two infinite sets equivalent sets? For example, is the set of whole numbers equivalent to the set of integers?" How would you respond?

4 A student asks, "How can we tell whether a set is finite or infinite?" How would you answer?

## Answers

1 One way of helping the student would be to use indirect proof as follows. Let A be an arbitrary set. Either $\phi \subseteq A$ or $\phi \nsubseteq A$. Suppose that $\phi \nsubseteq A$. It would then follow that $\phi$ must contain at least one element that is not in A. However, if $\phi$ contains at least one element, then $\phi$ is not the empty set. Thus $\phi$ must be a subset of A.

An intuitive way of helping the student follows. Let A be an arbitrary set containing say n elements. Then one way of getting a subset of A could be to take elements out of A one by one. After we take all of the elements out of A, we are left with the empty set, which would also be a subset of A.

2 No. Consider A where $A = \{x | 0 \leqslant x \leqslant 1, x \in \text{real numbers}\}$. This set has a largest element, namely 1, but it certainly is not finite.

This student is also assuming that there is some "natural" order associated with *every set.* It should be made clear to the student that this is *not* the case. For example, consider the set of students in the class. One might order this set according to height or according to weight, but one should be able to consider this set without reference to any such order.

3 Two infinite sets are not necessarily equivalent (i.e., contain the same number of elements). For example, the set of natural numbers is not equivalent to the set of real numbers. The proof below is standard in illustrating the above statement.

Assume that there is a one-to-one correspondence between the real numbers and the natural numbers. Let each real number be represented by a nonterminating decimal in this correspondence. If the following represents the correspondence between the natural numbers and the real numbers, then one can find a real number that is not listed.

*TABLE 2.12.1*

| *Natural numbers* | *Real numbers* |
|---|---|
| 1 | $0.a_{11}a_{12}a_{13}a_{14}\cdots$ |
| 2 | $0.a_{21}a_{22}a_{23}a_{24}\cdots$ |
| 3 | $0.a_{31}a_{32}a_{33}a_{34}\cdots$ |
| 4 | $0.a_{41}a_{42}a_{43}a_{44}\cdots$ |
| 5 | $0.a_{51}a_{52}a_{53}a_{54}\cdots$ |
| 6 | |
| ⋮ | |

Replace each digit along the diagonal, $a_{11}$, $a_{22}$, $a_{33}$, $a_{44}$, ..., by a different one, $c_1$, $c_2$, $c_3$, $c_4$, ..., where $c_n$ is not 0 or 9. The number given by the nonterminating decimal $0.c_1c_2c_3$ ... is a real number that is not paired with any natural number since it differs from each given number by at least one digit. Thus, our assumption that there was a one-to-one correspondence between the natural numbers and the real numbers is false.

4 A set is finite if we can count the elements in the set and the counting comes to an end. A set is infinite if in trying to count the elements in the set, the counting never ends.

More precisely, a set A is finite if there exists a nonnegative integer k such that the set A contains precisely k elements. That is, there is a one-to-one correspondence between the elements of A and the elements of the set $\{1, 2, 3, ..., k\}$ where $k \geqslant 1$. The empty set is considered to be finite since it contains precisely zero elements. A set is infinite if it is not finite. That is, for every positive integer n, one can find n + 1 distinct elements in A.

## 2.17 Simultaneous Systems of Equations

### Questions

1 A student is trying to solve the set of simultaneous linear equations below.

$$\begin{aligned} 2x + 3y &= 6 \\ \underline{-6x - 9y} &\underline{= -14} \end{aligned}$$

He comes up with 0 = 4 but doesn't know what it means. How would you help this student?

2 A student is trying to solve the set of simultaneous linear equations below.

$$\begin{aligned} 2x + 3y &= 6 \\ \underline{-4x - 6y} &\underline{= -12} \end{aligned}$$

A student ends up with 0 = 0 and doesn't know what to do from here. How would you help this student?

3 A student hands in the following work.

$$\begin{aligned} 2x + 3y &= 4 \\ \underline{x - y} &\underline{= 5} \\ -\frac{5}{2}y &= 3 \end{aligned} \qquad \begin{aligned} -x - \frac{3}{2}y &= -2 \\ \underline{x - y} &\underline{= 5} \\ y &= -\frac{6}{5} \end{aligned}$$

Is the student's work correct?

4 While studying a unit on systems of open sentences in two variables, a student asks, "What is the difference between equivalent systems of equations and equivalent equations?" What is your response?

## Answers

1 In solving the simultaneous linear equations.

$$\begin{aligned} 2x + 3y &= 6 \\ -6x - 9y &= -14 \end{aligned}$$

we are looking for the set of all ordered pairs (x, y) that satisfy both equations. Now

$$\text{(a)} \quad \left.\begin{aligned} 2x + 3y &= 6 \\ -6x - 9y &= -14 \end{aligned}\right\} \quad \text{is equivalent to} \quad \text{(b)} \quad \left.\begin{aligned} 6x + 9y &= 18 \\ -6x - 9y &= -14 \end{aligned}\right\}$$

$$\text{which is equivalent to} \quad \text{(c)} \quad \left.\begin{aligned} 0 \cdot x + 0 \cdot y &= 4 \\ -6x - 9y &= -14 \end{aligned}\right\} \quad \begin{aligned} (1) \\ (2) \end{aligned}$$

Since no ordered pair (x, y) is a solution of (1), then (c) has no solution. Since (a) and (c) are equivalent systems of equations, then (a) also has no solution. If one sketches the graph of (a), then the graph would consist of two parallel lines, which do not intersect and thus have no common solutions for the given equations.

2 In solving the simultaneous linear equations

$$\begin{aligned} 2x + 3y &= 6 \\ -4x - 6y &= -12, \end{aligned}$$

we are looking for the set of all ordered pairs (x, y) that satisfy each equation. Now

$$\text{(a)} \quad \left.\begin{aligned} 2x + 3y &= 6 \\ -4x - 6y &= -12 \end{aligned}\right\} \quad \text{is equivalent to} \quad \text{(b)} \quad \left.\begin{aligned} 4x + 6y &= 12 \\ -4x - 6y &= -12 \end{aligned}\right\}$$

$$\text{which is equivalent to} \quad \text{(c)} \quad \begin{aligned} 0 \cdot x + 0 \cdot y &= 0 \\ -4x - 6y &= -12 \end{aligned} \quad \begin{aligned} (1) \\ (2) \end{aligned}$$

Since any ordered pair (x, y) satisfies (1), then the solution set of (c) consists of all those ordered pairs (x, y) that satisfy (2), which is an infinite number since one simply chooses an x and then finds y. Since (c) is equivalent to (a), then the solution set of (a) is an infinite set and consists of all ordered pairs that satisfy (2). If one sketches the graph of (a), then the graph would turn out to be a straight line. The equations in (a) are equivalent equations and thus each has the same line for its graph.

3 The student is saying that the system of equations

$$\left.\begin{aligned} 2x + 3y &= 4 \\ x - y &= 5 \end{aligned}\right\}$$

is equivalent to $y = -6/5$, which is of course incorrect. The student should complete the problem, getting the equivalent system of equations as follows:

$$y = -6/5$$
$$x = 19/5.$$

Thus $\{19/5, -6/5\}$ is the solution set of the original system of equations.

4 Equivalent equations are equations that have the same solution set; for example,

$2x + 3 = 5$ is equivalent to $2x = 2$ or $x = 1$.

Equivalent systems of equations are *systems* (two or more) of equations that have the same solution set; for example,

$$\left.\begin{aligned} x + 2y &= 8 \\ 2x - 3y &= 2 \end{aligned}\right\} \text{ is equivalent to } \left.\begin{aligned} x &= 4 \\ y &= 2 \end{aligned}\right\}$$

Thus the concept of equivalent systems of equations is a generalization of equivalent equations. Notice, however, that even though equivalent systems of equations have the same solution set, the equations forming the equivalent systems need not be equivalent; i.e., $x + 2y = 8$ is not equivalent to $x = 4$, and $2x - 3y = 2$ is not equivalent to $y = 2$.

## 2.18 Slope

### Questions

1 "Can you determine the slope of a line from its graph if you don't know the equation of the line and vice versa." How would you answer this student?

2 A student asks, "Since slope is such an important concept, why hasn't the slope of a vertical line been defined?" What is your reply?

3 A student asks, "Why is the letter *m* used to denote the slope of a line?" How do you respond?

### Answers

1 Yes, you can. Consider the graph shown in Figure 2.18.1.

*FIGURE 2.18.1*

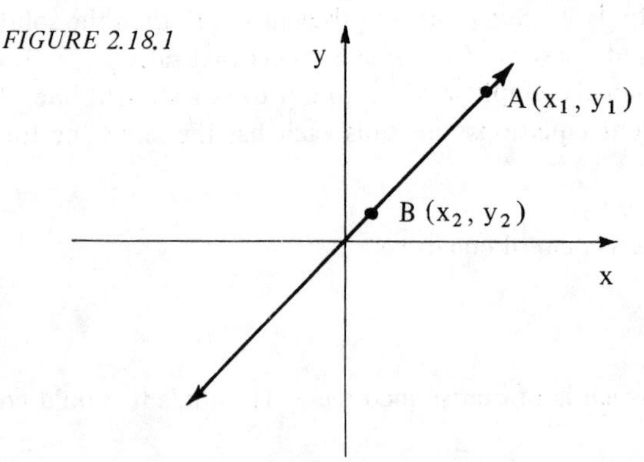

If we can pick convenient points such as A and B whose coordinates can be determined, then $\frac{y_1 - y_2}{x_1 - x_2}$ would be the slope of the line.

If the equation of the line is given, then two points can readily be determined by substituting into the equation, and the slope can be found as above.

2 Slope has been defined in a meaningful way as the change of the y coordinates of two distinct points, $\Delta y$, divided by the change of x coordinates, $\Delta x$, of these two points. When the line is vertical, then we run into the expression $\Delta y/\Delta x$ where $\Delta x = 0$ and $\Delta y \neq 0$. Since this expression does not represent any definite number, then we say that the slope of a vertical line is not defined as a number.

3 As far as we know there does not seem to be a good reason for using the letter *m* for slope. As Howard W. Eves points out in his book *Mathematical Circles Revisited,** "When lecturing before an analytic geometry class during the early part of the course, one may say: 'We designate the slope of a line by *m*, because the word slope starts with the letter *m*; I know no better reason',"

In many situations such as this one, we really do not know why a particular letter or symbol is used, such as $\theta$ for angle. However, what usually happens is that after a choice is made, this choice seems to carry over in the writings of other people. In this particular case, it might have been that since slopes were first studied with respect to mountains that the first letter of the word mountain was used to designate slope, but we really do not know.

*Howard W. Eves, *Mathematical Circles Revisited,* Prindle, Weber & Schmidt, Boston, Ma., 1971, p. 141, 267°.

## 2.19 Word Problems

### Questions

1 The problem is: "John rows up the river at 6 mph and down the river at 8 mph. What would John's rate be in still water and what is the rate of the current of the river?" A student says, "To solve this, I add 6 mph and 8 mph and divide by 2. This is John's rate in still water. To find the rate of the current, I subtract 7 from 8 or else subtract 6 from 7. They both give the same answer." Is the student correct?

2 The following problem and solution are done in class. "A person invests half a sum at 3.5% interest rate and half a sum at 4.5% interest rate. At the end of a year the person received \$300.00 in interest. How much did the person invest at each sum?"

$$\begin{aligned} \tfrac{1}{2}(.035)x + \tfrac{1}{2}(.045)x &= 300 \\ .0175x + .0225x &= 300 \\ .04x &= 300 \\ x &= \$7{,}500 \end{aligned}$$

A student asks, "Why can't you always average the two interest rates and then divide that percent rate into the interest to find the principal?" How do you reply?

3 The following problem is given to your class. "A pump can fill a tank in 3 hours. With the drain open, it will empty in 5 hours. With the pump working and the drain open, how long will it take to fill the tank?" A student handed in the following response.

$$\left(\frac{1}{3} - \frac{1}{5}\right)x = 1$$

$$\left(\frac{5}{15} - \frac{3}{15}\right)x = \frac{15}{15}$$

$$\frac{2}{15}x = \frac{15}{15}$$

$$x = 7\frac{1}{2}$$

Is the method correct? Is the solution correct?

4 The following problem is given to your class. "In an election one candidate received 96 votes more than his opponent. Since there were 804 votes cast, the candidates received___votes and___votes." A student says, "Divide 804 by 2 since there were two candidates. 804/2 = 402. One candidate got 402 + 96 = 498 votes. The other got 804 – 498 = 306 votes." A student asks, "Why doesn't this work?" How do you respond?

5 A ninth grade student is trying to solve the following word problem. "A boat travels downstream 24 miles in two hours and comes back upstream in four hours. Find the rate of the boat in still water." The student reasons that the current going one way cancels it going the other way. Therefore, the boat goes a distance of 48 miles in six hours, and the rate of the boat is 8 mph. Is the student correct?

6 The problem given is: "If 4 pounds of meat must cook for 2.5 hours, how much meat do you have if it must cook for 9 hours?" Student solution follows.

```
    .625
 4 |2.500
    2 4
     10
      8
     20
     20
```

```
  .625
 ×   9
 5.595
```

Is this student correct?

7 While studying word problems in Algebra I, a student in your class makes the following comment: "It seems to be the case that every integer can be expressed as a sum of consecutive integers except for those integers that are a power of two.

For example:

$$3 = 1 + 2$$
$$5 = 2 + 3$$
$$6 = 1 + 2 + 3$$
$$9 = 4 + 5$$

Is it true?"

What response do you make?

8 The following problem was given to your class. "Mike is 30 years older than Jim. Two years from now Mike will be twice as old as Jim. How old is Mike now?" A student solves as follows:

$x$ = Jim's age
$x + 30$ = Mike's age
$2x + 2 = x + 30$
$x = 28$ (Jim)
$x + 30 = 58$ (Mike)

Is the answer correct? Is the method correct?

9 Your class is doing a word problem concerning three consecutive even integers. You ask a student how can these numbers be represented and he replies, "...x, x + 2, x + 4." Is the student correct?

## Answers

1 Yes, the student is correct. Solving this problem using a standard method, one gets the following:

$x$ = rate of boat in still water,
$r$ = rate of current.

$$x + r = 8$$
$$x - r = 6$$
$$2x = 14$$
$$x = 7$$

Thus $r = 1$.

2 As the first equation shows, the student is correct only when half the money is invested at each interest rate for the same amount of time.

If the money is invested for different time periods or if some other combination other than half of the money at each interest rate is given, then the student's method will not work. However, a method might be worked out in the latter cases by using a weighted average.

3 The method is correct, and the solution would be completely correct if the student had put 7½ *hours* for the answer.

4 By adding 96 to 402 and subtracting 96 from 402, the student is saying that one candidate received 192 votes more than his opponent. If the student subtracts 48 (½ of 96) from 402 and then adds 48 to 402, the difference between the number of votes each candidate received would be 96.

5 Although the student shows good intuitive thinking in this problem, the student is incorrect. The main difficulty is that the current going one way does not cancel it going the other way, since the time going each way is not the same. If one solves the problem using the standard method, one gets the following:

$x$ = rate of boat in still water,
$r$ = rate of current.

$$24 = (x + r)2$$
$$24 = (x - r)4$$

Thus
$$12 = x + r$$
$$6 = x - r$$
$$x = 9$$
$$r = 3.$$

6 No, the student is not correct. In dividing 2.5 by 4, the answer of .625 represents the time needed to cook one pound of meat. In multiplying .625 by 9, the student is multiplying two numbers that both represent time. A correct way of doing the problem is:

```
     1.6
2.5 )4.00
     2 5
     1 50
```

Thus 1.6 represents the number of pounds of meat cooked in one hour. Then

```
  1.6
    9
 14.4
```

represents the number of pounds of meat cooked in 9 hours.

7 Yes, the student is correct. The student has conjectured this well-known result in number theory: The number of different ways of writing n as a sum of consecutive integers equals d(m) where m is the largest odd number dividing m and d(m) is the number of positive divisors of m*. For example, 15 has four odd divisors, 1, 3, 5, 15, and thus it can be written in four different ways as a sum of consecutive integers; namely,

$$15 = 7 + 8$$
$$15 = 15$$
$$15 = 4 + 5 + 6$$
$$15 = 1 + 2 + 3 + 4 + 5.$$

*Andrews, George, *Number Theory*, W. B. Saunders and Co., Philadelphia, 1971, p. 84, Problem 4.

The result is one that a teacher, using an inductive discovery teaching method, could use to good advantage in any class. For example, the teacher could give each student in the class five different numbers and see if each of those numbers that is not a power of two can be written as a sum of consecutive integers. After doing this, the teacher could ask the students to find the number of different ways each of their numbers can be written as a sum of consecutive integers. Perhaps after doing this, the students could conjecture the result given above. This result could be very useful in classes called general mathematics or basic skill classes. Notice the number of different addition combinations that a particular student has to perform while working on this problem.

We suggest that you do not ask your students to prove this result but rather have them conjecture the general result.

For more information on this problem, read:

Robert W. Prielipp and Norbert J. Kuenzi, "Sums of Consecutive Positive Integers," *The Mathematics Teacher*, Jan. 1975, pp. 18-21.

8 The student's answer is correct. If one compares the student's method with the following "standard" method of solving this problem, one might suspect that the student's method is incorrect and just happened to work in this case.

| | |
|---|---|
| x = Jim's age now | x + 2 = Jim's age in 2 years |
| x + 30 = Mike's age now | x + 32 = Mike's age in 2 years |

$$\therefore x + 32 = 2(x + 2)$$

$$x + 32 = 2x + 4 \quad (1)$$

However, if one compares (1) with the student's equation at this point, one finds that they are equivalent. If the problem was changed from "2 years from now" to "y years from now," it looks as if the student would get the following:

$$2x + y = x + 30 \quad (2)$$

Solving this generalized problem in the standard way, one gets

$$x + 30 + y = 2(x + y),$$

$$x + 30 = 2x + y \quad (3)$$

Thus even in the generalized problem, the student gets the correct equation. Whether the student just happened to stumble across this method by sheer accident and it happens to work in this problem, or whether the student has insight into the problem (or simplified his equation in his head) cannot be determined by what is given.

9 Yes, the student is correct. Many students and teachers assume that three consecutive even integers must be represented by 2y, 2y + 2, 2y + 4. However, if we let 2y = x, then it is obvious that either representation can be used.

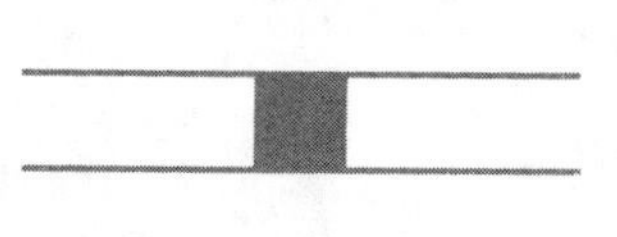

# 3. Geometry

## 3.1 Angles

### Questions

1 A student asks you if a dihedral angle is really an angle. What is your response?

2 One of your students says that the angle shown in (b) is bigger than the one in (a). How do you respond?

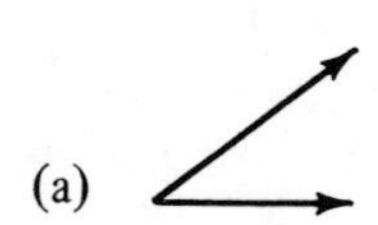


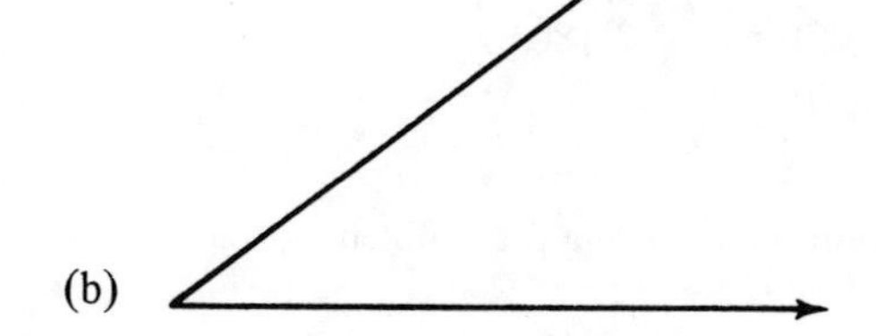


3 A student makes the statement, "In the diagram below a + b + c + d = 180 since the four angles make up a straight line." What is your reply?

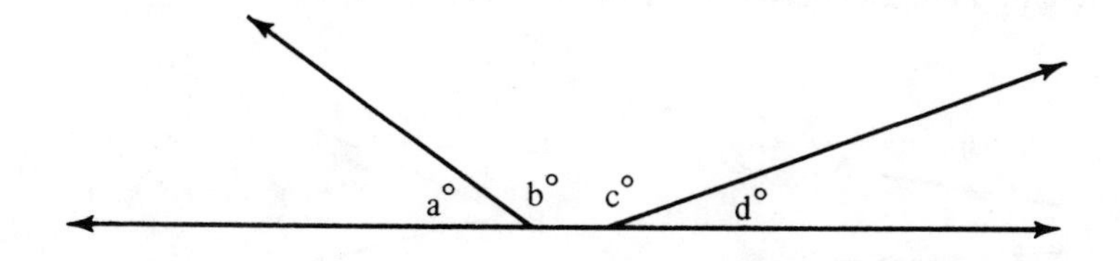


4 After showing students how to bisect an angle, how do you answer the student who asks, "Is it possible to divide an angle into three equal parts?"

5 A student says, "I think that the theorem which states that vertical angles are congruent is wrong. Look at the diagrams below. Angle 1 is not congruent to angle 3." How do you respond?

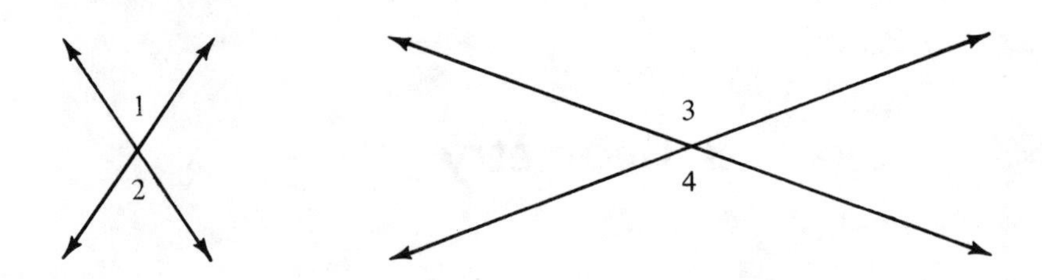


6 You draw the following figure on the blackboard.

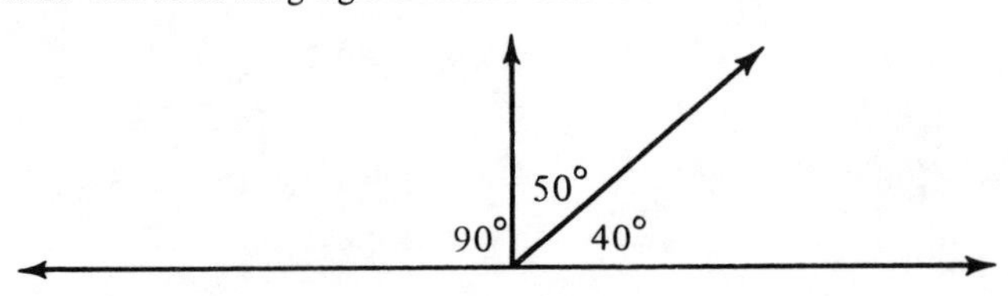


A student said that these angles are supplementary because their sum is 180°. Is the student correct?

7 The problem is: "The measure of each interior angle of an equiangular polygon is five times as large as the measure of an exterior angle of the polygon. How many sides does the polygon have?" Student's solution:

$$\overset{2}{5(\cancel{360})} = (n-2)\cancel{180}$$
$$10 = n - 2$$
$$12 = n$$

Is this method of solving the problem correct?

8 A student asks, "We have defined an angle as the union of two distinct rays that have the same endpoint but do not lie in the same line. However, in the figures below the angles are the same (i.e., same set of points rotate $\overrightarrow{BC}$ about point B) but they look different. I don't understand." How do you respond?

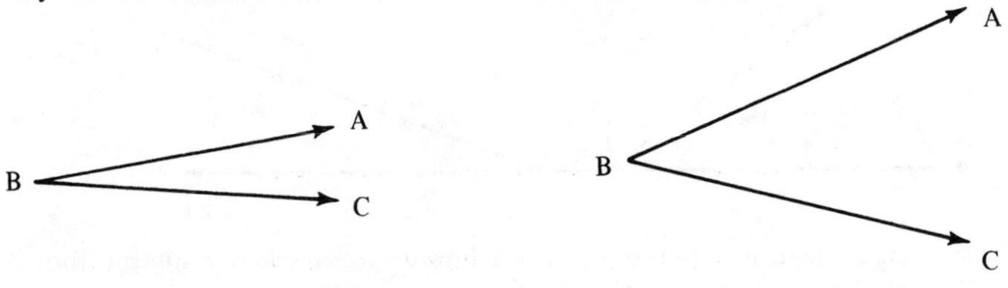


9 A student asks, "We said that a bisector of an angle is the set of all points that are the same distance from both sides of an angle. Yet in the figure below, R is not the same distance as S from the sides of the angle. What is wrong?" How do you respond?

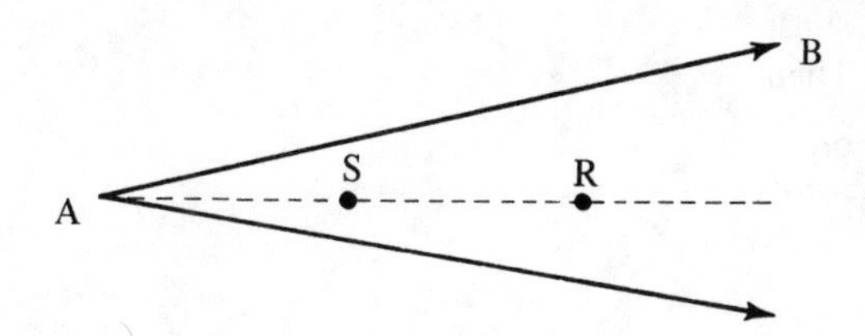


10 A student claims that there are four different angles in this drawing: ∠1, ∠2, ∠3, ∠4. What is your response?

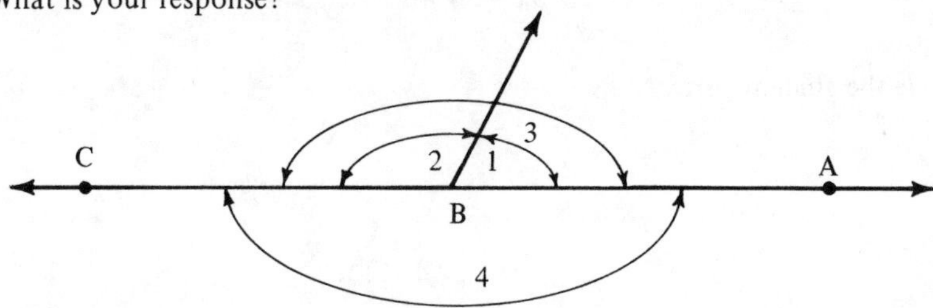


11 "In studying the concept of an angle, we have seen that it is related to an arc of a circle. Is there an analogue of this relationship in three dimensions?" How do you reply to this student?

12 A student asks, "Can two angles have a common side and a common vertex and not be adjacent?" How do you respond?

13 You are going over the following problem. "Represent as a single fraction, the measure of the complement of an angle whose measure is $x - 10$." One of the students says that he did the problem as follows:

Let $x$ = the measure of the complement of the angle
$x - 10$ = the measure of the angle

$$x + (x - 10) = 90$$
$$2x - 10 = 90$$
$$2x = 100$$
$$x = 50, \quad x - 10 = 40$$

Is the student correct?

14 A textbook makes the statement that an acute angle is an angle whose measure is less than 90°. A student then asks you if an angle whose measure is zero is an acute angle? How do you respond?

15 You have assigned the following problem to your class. "The measures of the acute angles of a right triangle have the ratio 3:4. Find, to the nearest whole minute, the number of degrees and minutes in the smaller angle." One student hands in the following work. Let $3x$ equal the measure of the smaller angle and $4x$ equal the meas-

ure of the larger angle. Then

$$7x = 90 \quad \text{or} \quad x = \frac{90}{7}.$$

```
   12.8
7 |89°60′
   7
   19            x = 12.8
   14           3x = 38°40′
    56
```

Is the student correct?

## Answers

1 A dihedral angle is really the 3-dimensional analogue of a planar angle. Thus a dihedral angle is not a special type of planar angle, but rather it is an "angle" in three dimensions. It is interesting to note that the measure of a dihedral angle is defined to be the measure of the plane angle formed on the faces of the dihedral angle by a plane perpendicular to the edge of the given dihedral angle.

For additional information, consult the references listed in Answer 11 of this section.

2 A good question to ask the student in this case is: "What do you mean by bigger?" What you should really be getting after is the distinction between a thing and its representation, in this particular case, an angle and its graphical representation.

One might ask the student which of the following is "bigger"

*3* or ***3***

This latter problem involves the difference between a numeral and a number and is the arithmetic counterpart to the given problem.

Essentially you want the student to see that each of the following is a representation for the same angle; i.e., ∠BAC = ∠DAE = ∠FAG, etc. (Figure 3.1.1).

*FIGURE 3.1.1*

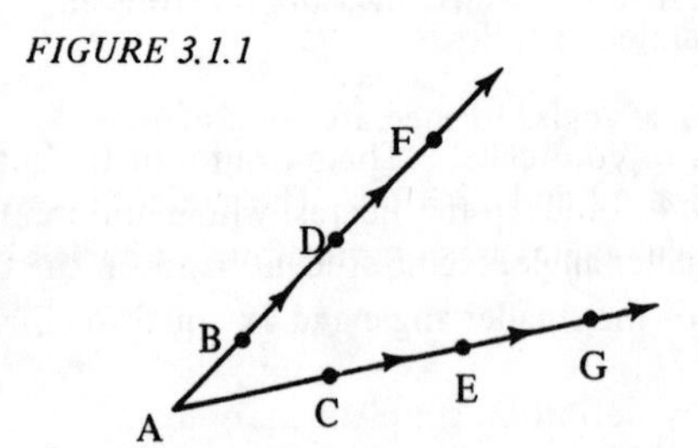

3 Suppose that we label the given diagram as shown in Figure 3.1.2.

*FIGURE 3.1.2*

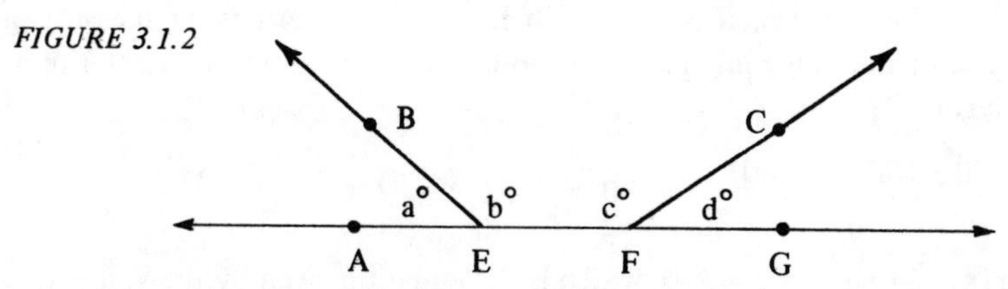


One way of answering this student is as follows. Suppose we separate Figure 3.1.3 as shown.

*FIGURE 3.1.3*

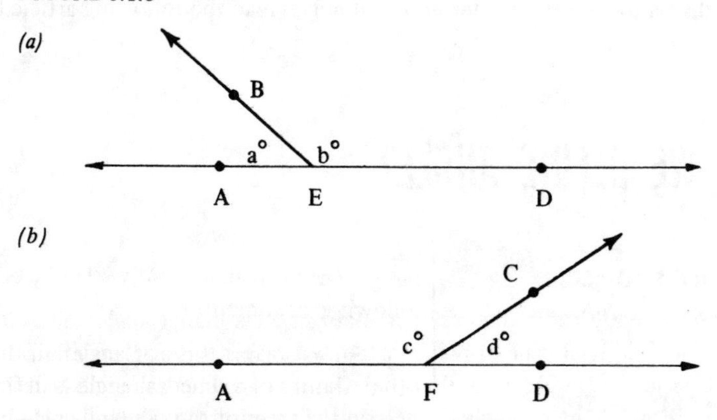


Doing this, one can see that m∠AEB + m∠BED = a + b = 180 and m∠AFC + m∠CFD = c + d = 180.

***Problem*** Under what conditions will the student's assertion be correct?

4 For some special angles, it is possible to divide an angle into three congruent parts, that is, to trisect an angle, but for the general case, one cannot trisect an angle using only a straightedge and compass. This involves one of the three famous problems from antiquity each of which cannot be solved by using only a straightedge and compass. The other two problems are constructing a square equal in area to a given circle and constructing the side of a cube whose volume is twice that of a cube of a given side. For more information on these problems consult:

Heinrich Tietze, *Famous Problems of Mathematics*, Graylock Press, New York, 1965.

5 The theorem that the student is using is stated as follows.

If two lines intersect, then the vertical angles formed are congruent.

Thus referrring to the given diagram, ∠1 ≅ ∠2 and ∠3 ≅ ∠4. The student is assuming that the theorem states that all vertical angles are congruent, which certainly is not true.

6 No, the student is not correct because by definition supplementary angles are *two* angles the sum of whose measures equals 180°.

7 Yes, the method is correct. The problem can be solved by letting n equal the number of sides of the equiangular polygon. Then $(n-2)\cdot(180)/n$ is the measure of an interior angle of the polygon. Likewise, $360/n$ is the measure of an exterior angle of this polygon. Thus

$$\frac{(n-2)\cdot 180}{n} = \frac{5(360)}{n} \qquad n \neq 0.$$

Therefore $(n-2)\cdot 180 = 5\cdot 360$, which is the equation given by the student.

8 In rotating $\overrightarrow{BC}$ about point B, the student does get a different set of points. Thus, the angles would not be the same; that is, they are not the same set of points. For an interesting discussion about the definition of angle, read the following article by Harry Sitomer and Howard F. Fehr.*

# HOW SHALL WE DEFINE ANGLE?

By **HARRY SITOMER**
Teachers College, Columbia University
New York, New York

and **HOWARD F. FEHR**
Teachers College, Columbia University
New York, New York

IN PUBLICATIONS of the School Mathematics Study Group, an angle is defined as the union of two noncollinear rays having the same origin. This definition is now widely used in the United States, and has found its way into many contemporary textbooks. On the other hand, many Europeans use the word "angle" to describe what SMSG calls an angular region: that is, the set of points in the union of the interior of an SMSG angle with the angle itself. The distinguished Professor Gustave Choquet of Faculty of Science, Institute Henri Poincaré, University of Paris, who has given considerable attention to the teaching of elementary geometry, maintains that the angular region approach to the topic is closer to pupils' intuition and hence is to be preferred pedagogically. He lists the following arguments:

1. Suppose one side of an angle is rotated about the vertex as center to another position. The pupil thinks he has the same set of points, two rays, but a "larger" or "smaller" angle. This can be, and is, terribly confusing.

2. Consider two angles with one common ray. If they are SMSG angles, their union is a set of three rays—a strange geometric animal. On the other hand, the union of the related angular regions is still an angular region.

3. Consider two angles having one ray and at least one interior point in common. Then as SMSG angles their intersection is one ray. The interesection of the related angular regions is still an angular region.

The last two arguments have the same force as "like should beget like."

4. The midray of an SMSG angle does not belong to the angle; however, it *is* a ray of the angular region if we accept the proposition that every interior point of the midray is in the angular region. There is no mathematical reason why a midray

*From *The Mathematics Teacher*, January 1967, pp. 18-19. Reprinted by permission.

of an angle must be a ray of the angle; but when we consider the analogous situation for segments, we find its midpoint to be a point of the segment. This disturbs the students.

This suggests a closer examination of the analogy between angular regions and line segments. They do resemble each other in a number of ways. Below is a list of parallel statements in which we have the following correspondences:

| | |
|---|---|
| Line segment | Angular region |
| Endpoint | Endray |
| Midpoint | Midray |
| Line measure | Angular measure |

1. The midpoint of a line segment belongs to the segment.

1′. The midray of an angular region belongs to the region.

2. The union of two collinear line segments having exactly one endpoint in common is a line segment.

2′. The union of two coplanar angular regions having one endray in common is an angular region.

3. The intersection of two coplanar line segments having one endpoint and one interior point in common is a line segment.

3′. The intersection of two coplanar angular regions having one endray and one interior ray in common is an angular region.

4. The measure of a line segment is accomplished by taking a piece of the segment, called a unit, and covering the segment with nonoverlapping (only one point in common) units. The number of covering units is the measure.

4′. The measure of an angular region is accomplished by taking a piece of angular region, called a unit, and covering the region with nonoverlapping (only one ray in common) units. The number of covering units is the measure.

Of course this analogy, like most analogies, cannot be pressed far without running into serious difficulty. The point of the matter is that we cannot even *begin* to make such analogies between line segments and the SMSG angles. A more fruitful analogy could be made between a segment, defined as a pair of points, and the SMSG angle. But for most students, the intuitive (natural) line segment is the convex set of two points and all the points between them—not a pair of points. Similarly, the angle should be the convex set formed by a pair of rays with common vertex and all rays (with the same vertex) lying between them—not a pair of rays.

There is also the relationship between circular arcs and angles that arises when we study angular measures and arc measures. When the vertex of the angle is the center of a circle, then the related arc is the intersection of the circle and the angular region; the related sector is the intersection of the circular region and the angular region. In neither case do we take or use the intersection with the SMSG angle.

In view of these cogent arguments, we in the United States should reconsider the definition we give to *angle*, especially at the elementary and secondary school level.

9 The student has misinterpreted the given statement. This reasoning error is common and is one that should always be anticipated. Essentially what you must do is interpret for the student what the statement means, i.e., that the perpendicular distance from S to $\overrightarrow{AB}$ equals the distance from S to $\overrightarrow{AC}$, and the distance from R to $\overrightarrow{AB}$ equals the distance from R to $\overrightarrow{AC}$. The statement does not mean that the distance from S to $\overrightarrow{AB}$ must be the same as the distance from R to $\overrightarrow{AB}$. In general, the point P is on the bisector $\overrightarrow{SR}$ iff the distance between P and $\overrightarrow{AB}$ equals the distance between P and $\overrightarrow{AC}$.

***Problem*** Prove the statement above that the point P is on the bisector $\overrightarrow{SR}$ iff the distance between P and $\overrightarrow{AB}$ equals the distance between P and $\overrightarrow{AC}$. Use the definition of an angle bisector: $\overrightarrow{SR}$ is the bisector of $\angle ABC$ if $m\angle RAB = m\angle RAC$.

10 The student is not correct. There are at least two different angles, $\angle 1$ and $\angle 2$. If a straight angle is defined as an angle in the student's text (some books do not define straight angle), then there are three different angles, $\angle 1$, $\angle 2$ and either $\angle 3$ or $\angle 4$. Angle 3 and angle 4 are the same angle; i.e., they are the same set of points.

11 For an answer to this question consult:

Howard Fehr, "On Teaching Dihedral Angle and Steradian" (Article 51), and "On Teaching Trihedral Angle and Solid Angle" (Article 52). *Mathematics in the Secondary Classroom: Selected Readings*, eds. Gerald Rising and Richard Wisen, Thomas Y. Crowell Co., New York, 1972. (These articles also may be found in the April 1958 and May 1958 issues of *The Mathematics Teacher.*)

12 Consider Figure 3.1.4. $\angle BAC$ and $\angle BAD$ are not adjacent even though they have a common side and a common vertex since adjacent angles do not have interior points in common.

*FIGURE 3.1.4*

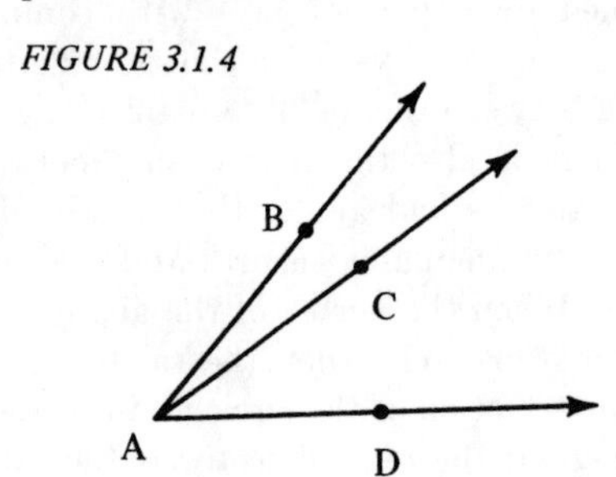


13 No, the student is not correct. Although this problem is easy for you to analyze, convincing some students that the method given is incorrect turns out to be quite difficult. Let y equal the measure of the complement of an angle whose measure is $x - 10$. Then

$$y + (x - 10) = 90.$$

Thus $y = 90 - (x - 10) = 100 - x$.

The student has made the false assumption that if $x - 10$ is the measure of an angle then x is the measure of its complement.

14 An angle in plane geometry is usually defined as the union of two noncollinear rays with the same endpoint. If an angle has a measure of zero, then these two rays coincide, and thus they are usually considered as one ray. Some definitions of angle exclude an angle of measure zero because of the above. However, in some texts and especially in trigonometry courses, an angle is defined differently (perhaps by including a rotation as part of the definition), and it is convenient to include an angle of measure zero.

15 No, the student is not correct. The correct answer is $x = 90°/7 = 12°52'$ and thus $3x = 38°36'$.

***Problem*** Why is the student's method of dividing ($7\overline{)89°60'}$) not correct?

## 3.2 Axioms and Definitions

### Questions

1 A student questions the statement that a point is an undefined term. The student reasons that a point can be defined as the intersection of two lines, and that a line could then be defined as the intersection of two planes. Thus, the only necessary undefined term in a high school geometry class would be *plane*. What is your response?

2 A student asks, "What is the difference between a postulate and an axiom?" How do you respond?

3 A student makes the following statement. "Explain how the completeness axiom for the real numbers provides the numbers necessary to fill up the number line, not already filled by the rationals." How do you respond?

4 A student comments that we often use function notation in other ways. For example, the notation that we use to name segments, rays, or lines, etc. is actually function notation. What would your reaction to this statement be? Explain what the student meant using the following symbolism.

(a) $\overleftrightarrow{AB}$ (b) $\angle ABC$

Can you show what the function symbol f refers to in each example and what f(x) refers to?

5 A student asks, "In geometry if the definition of the word *point* is undefined, how is it possible to use something that is not really there?" How do you respond?

6 "What is the difference between *necessary* and *sufficient* in the statement and proof of a theorem?" How do you respond to this student?

7 A student asks, "If $AB \cong CD$ means $AB = CD$, why do we need the notation $AB \cong CD$ since we have the notation $AB = CD$ already?" How do you respond?

8 A student asks you why axioms are not proved as theorems are. How do you respond?

## Answers

1 For an answer to this question, read the following two letters, which appeared in *The Mathematics Teacher*.*

DEAR EDITOR:

In a recent discussion in a high school geometry class, one of my students questioned my statement that a point is an undefined term. His reasoning was that a point can be defined as the intersection of two lines. And then I carried his reasoning a little further and said that a line could then be defined as the intersection of two planes. Thus, the only necessary undefined term in a high school geometry class would be "plane."

Would you tell me if there is an error in our reasoning that we have overlooked? Or is this a good way to teach high school geometry—with only one undefined term, rather than three?

Sincerely yours,
JO-ANN SCHIFFBAUER
*Lake High School*
*Hartville, Ohio*

DEAR MRS. SCHIFFBAUER:

Your students should be commended for their interest and curiosity concerning the structure of their geometry. The possibility of reducing the number of undefined terms is a most intriguing one. If, however, we attempt to define line and point as your class suggests, we might find ourselves in logical difficulties.

Suppose we define a point as the intersection of two lines. Since we are speaking of a set theoretic intersection, we would be implying that we already knew something about points, i.e., lines are sets of points. A similar difficulty would arise concerning the proposed definition of line.

A more reasonable approach might be to leave plane and point as undefined. Then state an axiom which indicates that a plane is a set of points and proceed from there. This would mean a complete re-examination of the axiomatic structure of high school geometry—a task of considerable magnitude.

*From *The Mathematics Teacher*, May 1970, p. 396. Reprinted by permission.

2 In older texts, such as Euclid's, there was a difference between these terms. A postulate was an assumption applicable only to a particular branch of mathematics. For instance, in geometry, there is the postulate that states: two points determine one and only one line. An axiom was an assumption applicable to all branches of mathematics; e.g., equals added to equals give equals. In modern texts, the distinction is not given, and the two terms are synonymous.

3 For an answer to this question, read the well-written section "The Completeness of the Real Number System" beginning on page 31 of:

William G. Chinn and N.E. Steenrod, *First Concepts of Topology*, Random House, New York, 1966.

4 The student should certainly be commended for good thinking.

(a) The notation $\overleftrightarrow{AB}$ can be related to function notation as follows:

*Domain* Pairs of points.

*Range* The geometric figure consisting of the points on line $\overleftrightarrow{AB}$; i.e., f(A, B) is the set of points on line $\overleftrightarrow{AB}$. The function symbol f refers to $\leftrightarrow$ and the rule is to associate the points on the line $\overleftrightarrow{AB}$ with the pair of points A and B.

(b) The notation $\angle ABC$ can be related to function notation as follows:

*Domain* Triples of points.

*Range* The geometric figure consisting of the points on the angle $\angle ABC$, f(A, B, C), is the set of points on $\angle ABC$.

The function symbol f refers to the angle and the rule is to associate the points on ∠ABC with ABC.

*Problem 1* If a student uses the notation ΔABC, what does he mean?

*Problem 2* Find three other examples in mathematics of the notion involved in Question 4.

5 The word *undefined* is not synonymous with *nonexistence*. This idea might be conveyed to the student by asking her for a definition of set. A student will usually come up with something like "A set is a group of objects." However, what is a group? After several exchanges of this nature, you can convince the student that set is also usually left as an undefined term. Does this mean that sets do not exist? It might also be pointed out to the student that many of the concepts that we study are not physical objects but are abstract ideas that exist only in our minds.

6 The terms *necessary* and *sufficient* are usually associated with theorems that are written in the form of a conditional or biconditional. For example in the statement, "If a triangle is equilateral, then it is isosceles," the statement, "A triangle is equilateral," is a condition that is sufficient or guarantees that a triangle is isosceles. The statements "the fact that a triangle is isosceles is necessary for a triangle to be equilateral" or "before a triangle can be equilateral, it must be isosceles," can be stated in another way: "If a triangle is not isosceles, then there is no way for it to be equilateral."

Perhaps an illustration that might be useful in conveying the difference between the terms necessary and sufficient would be the following. Consider a bug that is traveling in a straight path such as on the number line below. Assume that the bug starts at point A and always travels in a straight path on the number line.

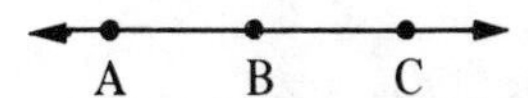


Suppose that our attention is directed away from the bug for a moment. If upon looking at the bug again, we observe that the bug is at point C, then under the conditions of the problem, we know that the bug must have passed through point B. That is, seeing the bug at C is sufficient for us to conclude that the bug must have passed point B. However, if we observe the bug at point B, then this informaion is not sufficient for us to say that the bug passed through C. However, we might say that before the bug can get to C, the bug has to pass at least once through B; i.e., it is necessary for the bug to pass through B before it can get to C.

7 The notation AB = CD means that the distance between points A and B is equal to the distance between the points C and D. The notation $\overline{AB} \cong \overline{CD}$ means that the segments $\overline{AB}$ and $\overline{CD}$ have the same size and shape. Thus, in the first case we are talking about the *distance* between points while in the second case we are talking about *a set of points* (the set of points contained in segments $\overline{AB}$ and $\overline{CD}$). It is not absolutely essential that the two different notations be used. In fact, at least one textbook series has used the notation $\overline{AB} = \overline{CD}$ to include both interpretations discussed above with the context of the problem determining which interpretation

was meant. However, in some cases, there might involve ambiguity of meaning, and thus it is now standard practice to use both notations.

8 In proving statements such as theorems in mathematics, one would have to use other statements or theorems or definitions as reasons in the proof. These statements used as reasons would likewise depend on other statements, theorems and definitions. This process would continue until one gets back to the very beginning or foundation of the subject. Eventually one has to stop in this process and the question comes up as to how these beginning statements are proved. In other words, there are no other theorems that one can use as reasons for proving these beginning statements and thus these beginning statements must be assumed to be true. These are the axioms of the system under consideration. Indeed mathematics is defined as the study of axiomatic systems. The process is one that is very similar to the idea of undefined terms in mathematics.

## 3.3 Circles and Constructions

### Questions

1 A student asks you if a circular region is really a region. How do you respond?

2 While studying circles in your geometry class, a student asks you what is the relationship between a tangent to a circle and the tangent of an angle that he studied in Algebra I. What is your answer?

3 A student asks you why great circles have the name *great*? How do you respond?

4 You ask your students to draw two circles such that they have three common tangents. One student says that it can't be done. Is the student right?

5 A student asks you if all circles are similar. How do you respond?

6 A student asks you if it is always possible to find a great circle through any two points in a given sphere. One student claims that it is impossible to find a great circle through the points P and Q in the diagram below. Is this student correct?

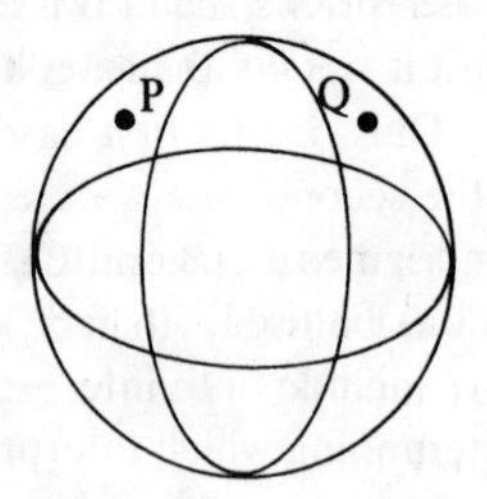

7 A student asks if you can draw a circle by using only straight lines. How do you respond?

8 "Can circles be circumscribed only about regular polygons?" How would you reply to this student?

9 A student asks you why the center of an arc is not also the midpoint of the arc? How do you respond?

10 "In the constructions that we have done, we were allowed to use both a compass and a straightedge. How many of these constructions can be done with only a compass?" How would you respond to this student?

## Answers

1 It depends on the definition of region that is used. One definition is as follows:

> A region is a set of points in a plane for which each of the following holds:
>
> (a) Any two points of the set are the endpoints of a polygonal path, which is contained entirely in the set.
>
> (b) Each point of the set is on at least two noncollinear line segments, which are contained in the set.

If this definition is used, then a circular region is a region.
In some texts a region is identified with a polygonal region and is usually defined as:

> A polygonal region is the union of a finite number of coplanar triangular regions, such that if any two of these intersect, the intersection is either a segment or a point.

If this definition is used, then one can say that a circular region is not a polygonal region. If polygonal region is synonymous with region, then clearly a circular region is not a region.

However, it should be pointed out that according to the definitions above the terms *polygonal region* and *region* are not synonymous. If a geometric figure is a polygonal region, then it is a region, but the converse is not valid. For example, a plane is a region, but it is not a polygonal region. It is usually the case, however, that one would want a circular region to be a region.

2 Consider Figure 3.3.1 where 0 is the unit circle, $x = 1$ is a line tangent to the circle and $\overrightarrow{OT}$ intersects the dashed line $x = 1$ at A. Now ray $\overrightarrow{OT}$ has slope $(\sin \phi)/(\cos \phi)$ or $\tan \phi$ and thus ray $\overrightarrow{OT}$ is contained in the line $y = (\tan \phi)x$. Thus the ray $\overrightarrow{OT}$ intersects the line $x = 1$ at the point $A(1, \tan \phi)$.

*FIGURE 3.3.1*

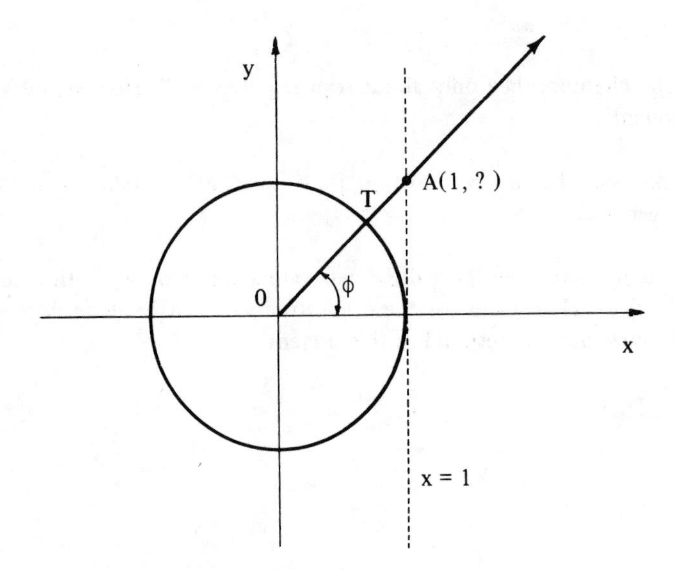


3 The term *great circle* probably originated because of its use in navigation; i.e., the shortest distance from one point on the earth to another point is along the path of a great circle. The adjective *great* was probably used because these circles have the largest circumference of any circles on a sphere.

4 No, the student is not correct. In Figure 3.3.2, $l_1$ is a common internal tangent while $l_2$ and $l_3$ are common external tangents.

*FIGURE 3.3.2*

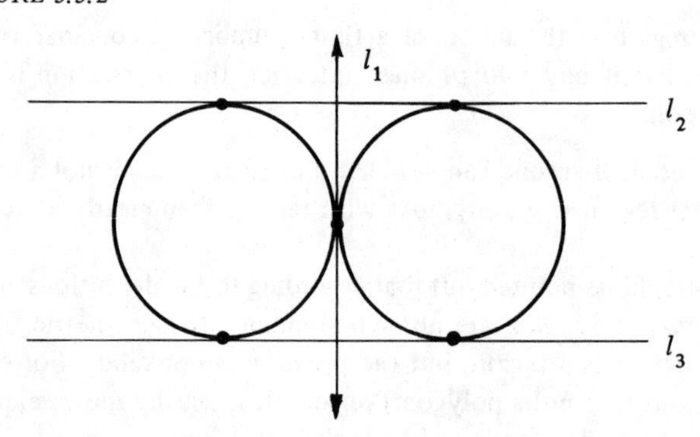


5 For an answer to this question, read:

Paul Johnson, "Are Circles Similar?", *The Mathematics Teacher*, January 1966, pp. 9-13.

6 Suppose one considers a sphere, as shown in Figure 3.3.3, with N representing the north pole and S, the south pole. If R is a point on the sphere, then it should be easy for the student to see that there is a great circle thru N and R. Simply rotate the great circle C, holding the points at N and S fixed, until C meets R. One can now apply the same technique to the sphere illustrated in this question. Simply rotate the sphere so that P corresponds to N, and then consider Q = R.

*FIGURE 3.3.3*

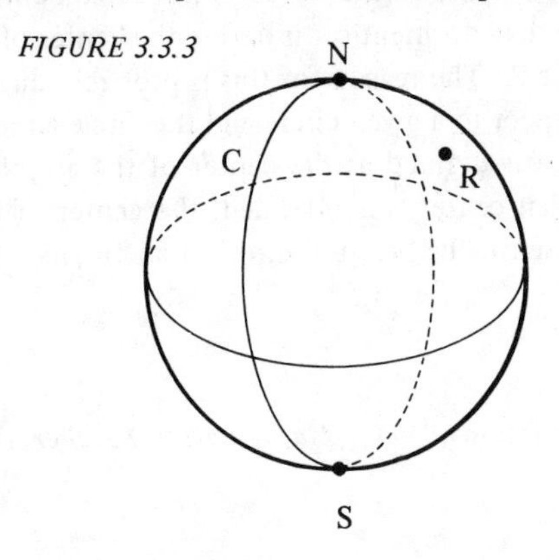


7 Yes, you can. One way of doing this is given below. Cut a circle out of wax paper. Any circle of radius r would be adequate, but the larger it is, the better. Locate the center of the circle and call it C. Lay the circle on a flat surface, and bring a point on its circumference up to meet point C; then crease. Repeat this process about thirty times, using different points on thę circumference. (About thirty repetitions should indicate the process to the student.) The creases will then enclose a circle.

***Problem 1*** How would you modify the above procedure in order to get an ellipse, a hyperbola, a parabola?

***Problem 2*** How would you convince the student that the process indicated and the processes you would use for Problem 1 do lead to the desired configurations?

***Problem 3*** If you have studied differential equations can you provide more insight into the above construction? That is, what idea is relevant to this question?

8 No. Consider the circle given and the quadrilateral ABCD shown in Figure 3.3.4. Obviously, quadrilaterial ABCD is not regular. However, for quadrilaterals, there is a condition involving angles that must be met in order for a circle to circumscribe it. Can you find out what it is?

*FIGURE 3.3.4*

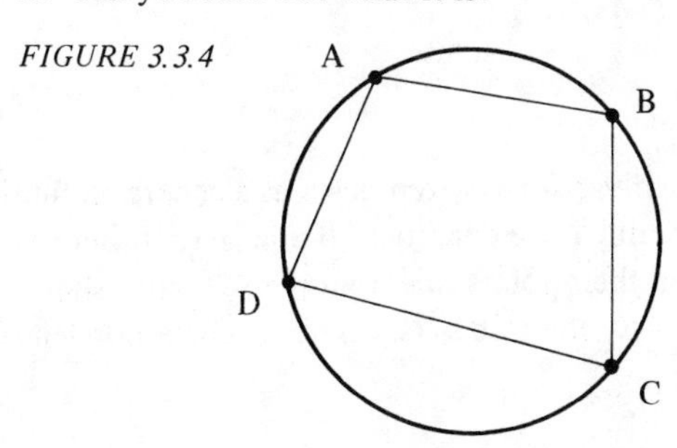

*Problem 1* In Question 8, we found out that the answer is no because of the cases involving quadrilaterals. If polygons of n sides, $n \geq 5$, are the only polygons considered, is the answer to the student's question still no? Explain.

*Problem 2* Can a circle be circumscribed about any triangle? Explain.

9 Many geometry textbooks used at the secondary school level do not even mention the idea of the center of an arc. Those that do mention it define the center of an arc of a circle P to be the center of circle P. The reason for this is probably due to the fact that an arc is described with respect to a given circle and the circle already has a center. If the midpoint of an arc was defined as the center of the arc, then possible confusion might result over which center was intended, the center of the arc or of the circle. Also, we have a name for the "centerpoint" of an arc, namely, the midpoint.

10 For an answer to this question, read:

J.H. Hlavaty, "Mascheroni Constructions," *The Mathematics Teacher*, November 1957, pp. 482-87.

## 3.4 Congruence

### Questions

1 A student asks, "I understand what is meant by two triangles being congruent, but how do we define congruence for sets of figures in general?" How do you respond?

2 A student asks, "In proving that two triangles are congruent, we have seen that it is not necessary to test the congruence of all six pairs of corresponding parts. Three pairs are sufficient, SSS or SAS or ASA. How many pairs would be sufficient for a quadrilateral or, in fact, any polygon?" What is your answer?

### Answers

1 The following statement or some such variant is often given as a general definition of congruence. Two sets of points X and Y are congruent if and only if there exists a one-to-one correspondence between their points such that if $r \in Y$ corresponds to $a \in X$ and $s \in Y$ corresponds to $b \in X$, then $\overline{ab} \cong \overline{rs}$. Such a correspondence is called an isometry.

2 It can be shown that for convex quadrilaterals, one need only have five corresponding pairs in order to insure congruence of the quadrilaterals. One of these corresponding pairs is four sides and one angle.

*Problem 1* Find all other corresponding pairs that would insure congruence of convex quadrilaterals.

*Problem 2* Find all other corresponding pairs that would insure congruence of pentagons and hexagons.

*Problem 3* Generalize your answer to Problem 2, for a geometric figure of n sides.

## 3.5 Convexity and Inequalities

### Questions

1 While studying convex polygons, a student asks, "Does the theorem concerning the sum of the interior angles of a polygon hold not only when it is convex but also when it is concave?" How would you answer the student? (*Hint* The sum of the interior angles of a convex polygon of n sides is $(n - 2)180°$.)

2 A student asks you if the empty set or a set containing one point are convex sets. The student does not think so because the textbook's definition states that for every two distinct points A and B, $\overline{AB}$ must be in the figure. How would you answer the student?

3 A student asks, "Are there situations in which an inequality is true in both half planes?" How would you answer?

4 "How do I know that the feasible region which arises in a linear programming problem will be a convex set?" How do you respond to this student?

### Answers

1 For an answer to this question, read the following article by R. Berman and M. Berman.*

*From *The Mathematics Teacher*, October 1963, pp. 403-406. Reprinted by permission.

# Concave polygons

ROSLYN M. BERMAN, *Cranston High School East, Cranston, Rhode Island.*
MARTIN BERMAN, *Physics Department, Brown University, Providence, Rhode Island.*

*Finding the sum of the interior and exterior angles of a concave polygon*

ALL PLANE GEOMETRY books examined by the present writers contain a discussion of convex polygons. Many texts define a polygon, convex polygon, and concave polygon, in that order. It is then explicitly stated that throughout the remainder of these texts the term "polygon" will signify a *convex polygon.*

Why do geometry books neglect to discuss concave polygons? Perhaps their authors are not aware that convex and concave polygons have many common properties, or they do not wish to confuse students with two different types.

Some high school geometry students asked one of the present writers whether the theorem concerning the sum of the interior angles of a polygon holds not only when the figure is convex, but also when it is concave. Consideration of this problem led to the following.

### THE SUM OF THE INTERIOR ANGLES OF A CONVEX POLYGON

First we will present proofs of how to determine the sum of the interior angles of a convex polygon. It will then be shown that these demonstrations cannot always be extended to a concave polygon.

*Method* 1. Euclid's *Elements*[1] is one of many subsequent texts which employed the means described in this paragraph. Figure 1 illustrates the convex polygon *ABCDEFG*. Pick any arbitrary vertex, such as *A*, and draw diagonals from *A* to all the other vertices. It is seen that if there

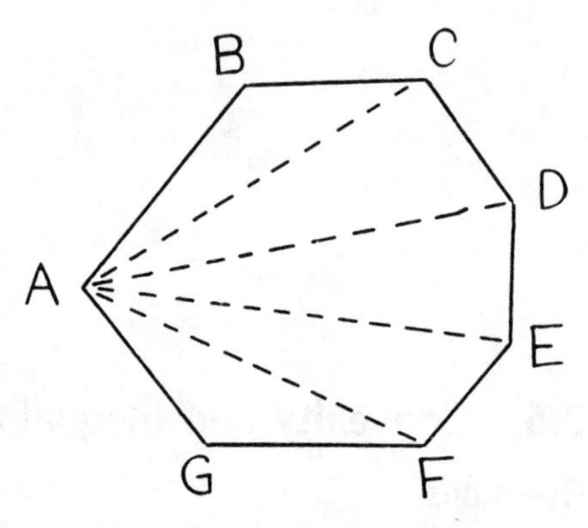


*Figure* 1

are $n$ sides to the polygon, $(n-2)$ triangles will be formed. The sum of the angles of the triangles will be equal to the sum of the interior angles of the polygon. The sum of the interior angles of a convex polygon of $n$ sides is therefore $(n-2)180°$.

*Method* 2. Any interior point $O$ of the polygon is chosen as the common vertex; lines are drawn from all the vertices of the polygon to point $O$ (Fig. 2). If the original

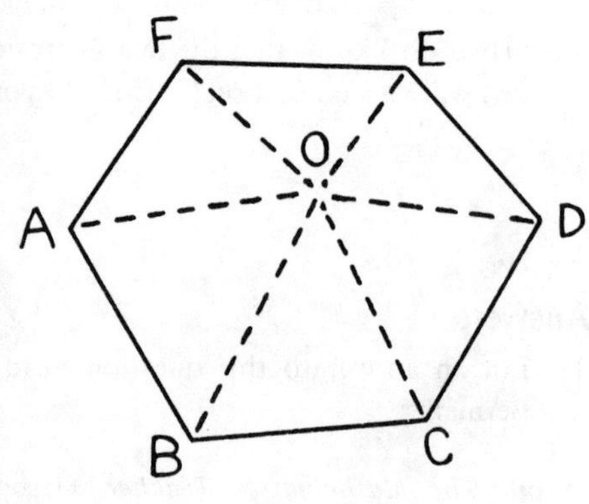


*Figure* 2

[1] Thomas L. Heath (trans.), *The Thirteen Books of Euclid's Elements* (Vol. 1. 3 vols.; second edition; New York: Dover Publications, Inc., 1956), p. 322.

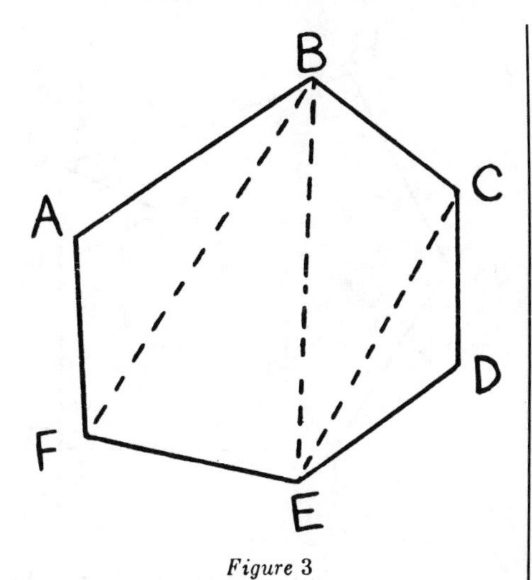

*Figure 3*

polygon has $n$ sides, $n$ triangles will be formed. Accordingly, the sum of the interior angles of an $n$-sided convex polygon is equal to the sum of the angles of the $n$ triangles minus the sum of the angles around point $O$. The sum of the angles around point $O$ is, of course, 360°. This again yields the correct answer.

*Method* 3. Diagonals are drawn so as to fill the inside of the polygon with separate adjacent triangles (Fig. 3). To accomplish this, none of the internal lines may cross one another. It can easily be seen that $(n-2)$ triangles will thereby be formed if the polygon has $n$ sides, once more indicating the proper conclusion.

To determine if the proofs given above hold for concave polygons, examine Figure 4. Polygon $ABCDEFGH$ possesses no vertex, as in Method 1, nor interior point,

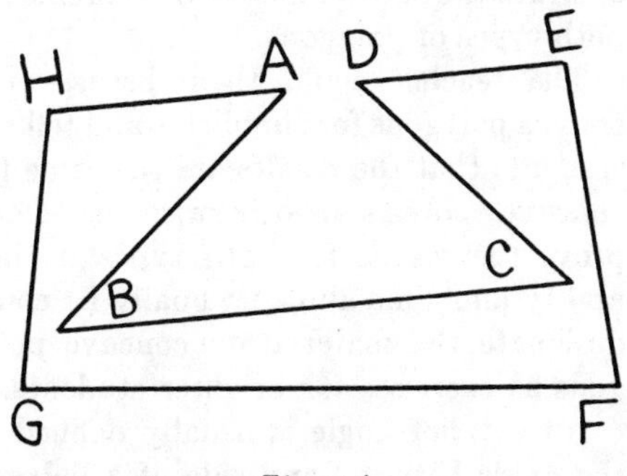

*Figure 4*

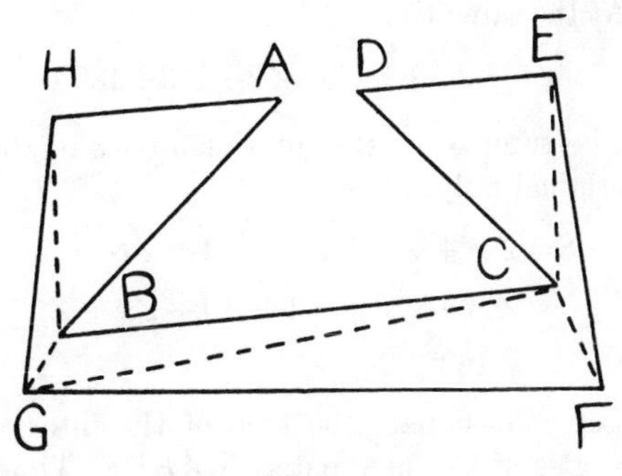

*Figure 5*

as in Method 2, from which triangles can be drawn whose sides will not cross the sides of the original polygon; many complications will ensue if such crossing does occur. As shown in Figure 5, Method 3 may be applied without difficulty. Method 3 is thus seen to be more general than either Methods 1 or 2. Although it seems intuitively obvious that Method 3 would hold for any polygon, the proof is difficult for intricate figures and, for this reason, was not attempted. Instead, the present writers have developed a simple means of proving that the sum of the interior angles of any $n$-sided polygon, be it convex or concave, is $(n-2)180°$.

### Equivalent $-m$ polygons

Consider a concave polygon which has only one concavity or re-entrant angle (an interior angle greater than 180°). In Figure 6, $\angle 6$ is the re-entrant angle. Construct line $BD$, thus forming $\triangle BCD$. $\angle 5$ is an exterior angle of $\triangle BCD$, so that

$$\angle 5 = \angle 2 + \angle 3.$$

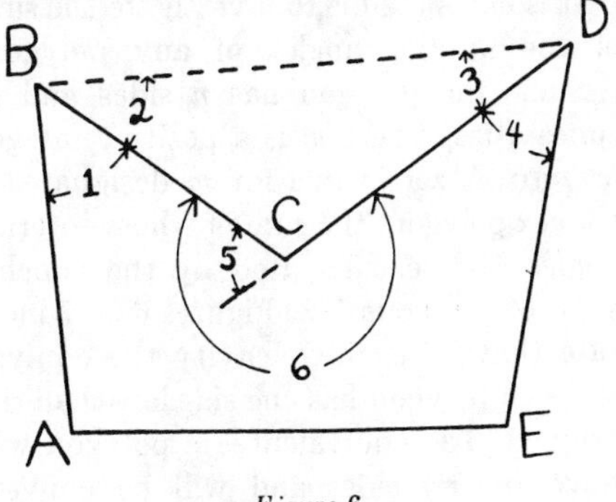

*Figure 6*

At the same time

$$\angle 6 = \angle 5 + 180° = \angle 2 + \angle 3 + 180°.$$

The sum, $S$, of the interior angles of the original polygon is

$$S = \angle A + \angle E + \angle 1 + \angle 4 + \angle 6$$
$$= \angle A + \angle E + \angle 1 + \angle 4 + \angle 2 + \angle 3 + 180°.$$

Let $S'$ designate the sum of the interior angles of the new polygon $ABDE$. Thus, the following is true:

$$S' = \angle A + \angle E + \angle 1 + \angle 2 + \angle 3 + \angle 4.$$

This gives the desired result, that

$$S' = S - 180°.$$

The new polygon is equivalent to the original polygon (with respect to the sum of its interior angles), except that the convex figure $ABDE$ contains 180° less. Since it is equivalent minus 1 times 180°, the present writers call it an equivalent$-1$ polygon to the original.

Suppose the concave polygon has more than one concavity, as in Figure 7. Then the construction described above is performed for each concavity. Polygon $ABCEFGI$ is equivalent$-2$ to polygon $ABCDEFGHI$, which means that it has the same total number of degrees in its interior angles minus 2 times 180°. If the polygon has $m$ concavities, the construction would be repeated $m$ times, resulting in an equivalent $-m$ convex polygon.

### THE SUM OF THE INTERIOR ANGLES OF ANY POLYGON

It is now possible to investigate the sum of the interior angles of any polygon. Assume the polygon has $n$ sides and $m$ concavities, where $m$ is a positive integer or zero. A zero-value for $m$ designates a convex polygon, the sum of whose interior angles can be calculated by the conclusions of Methods 1–3. Figures 6 or 7 indicate that for each concavity the equivalent$-m$ polygon has one side less than the original. The equivalent$-m$ polygon will have $(n-m)$ sides and will be convex.

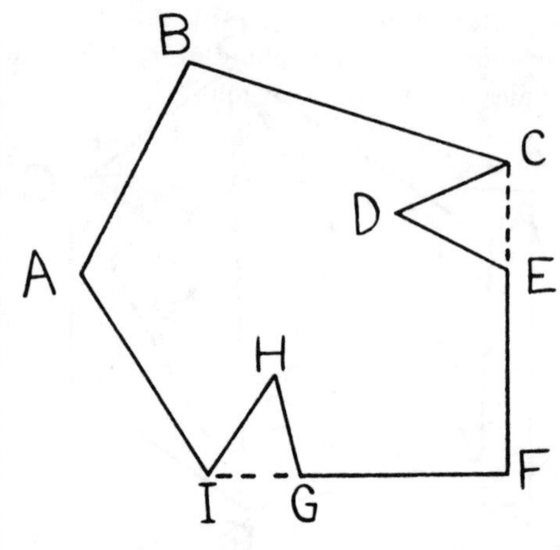


*Figure 7*

From the previous sections we know the following:

$$S = S' + 180°m$$

and

$$S' = [(n-m)-2]180°.$$

Combining the above gives

$$S = [(n-m)-2]180° + 180°m$$
$$= (n-2)180°.$$

### THE SUM OF THE EXTERIOR ANGLES OF A CONCAVE POLYGON

A topic usually considered with regard to the convex polygon, aside from the sum of its interior angles, is the total measure of its exterior angles. It is questioned: Can the concept of exterior angles, as defined for triangles and convex polygons, be extended to concave polygons? If so, it would be pedagogically desirable to demonstrate the same angle-sum theorems for both types of polygons.

The teacher could then discuss only convex polygons for simplicity and tell the students that the results are the same for concave polygons; otherwise he could prove the results for both types if class ability and time allow it; finally he could designate the material on concave polygons as exercises for brighter students.

An exterior angle is usually defined as the angle between any side of a polygon extended and the adjacent side not extended. This is the definition used for triangles and convex polygons.

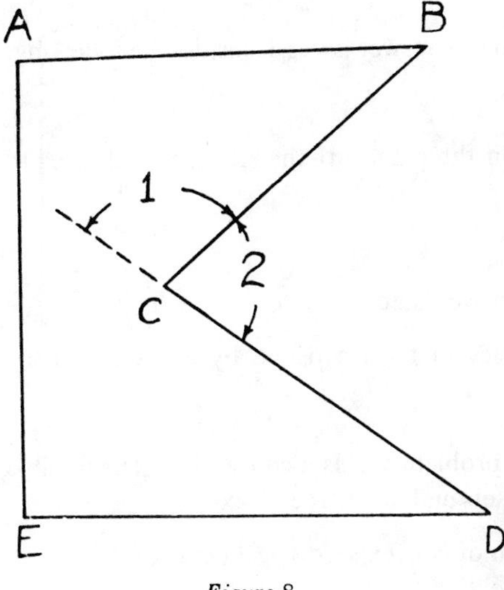


*Figure* 8

Consider the exterior angle of a reentrant angle of concave polygon $ABCDE$ (see Fig. 8). Which is the exterior angle of interior angle $BCD$? Is it $\angle 1$ which would fit the definition but is inside the figure (which seems strange for an exterior angle) or is it $\angle 2$, which is exterior to the figure but does not seem to fit the definition? To decide this question, it will be necessary to calculate the sum of the exterior angles and see which definition will give results consistent with the definition of convex polygons.

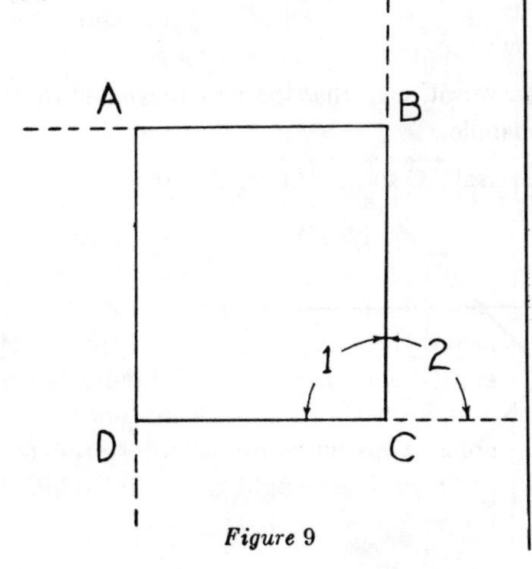


*Figure* 9

To calculate the sum of the exterior angles of a convex polygon, look at Figure 9. $\angle 1$ is the interior angle and $\angle 2$ is its corresponding exterior angle. It is seen that the sum of an interior angle and its corresponding exterior angle is 180°. Thus, for a convex polygon of $n$ sides, the sum of the exterior and interior angles is $180n°$. However, the sum of the interior angles is $(n-2)180°$ and we have:

$$\text{(sum of exterior angles)} + (n-2)180° = 180n°.$$

This gives the well known result that the sum of the exterior angles of a convex polygon is 360°.

We would like the sum of the exterior angles of a concave polygon also to be 360°. In the above proof, the reasoning depended upon the supplementary angle relation between each interior angle and its corresponding exterior angle. A reentrant angle is, by definition, greater than 180°. In order that the sum of it and its exterior angle be 180°, something must be subtracted from the interior angle. Looking at Figure 8, if $\angle 1$ is subtracted from interior angle $BCD$ we are left with 180°. We are thus led to the following definition of an exterior angle of any polygon: *An exterior angle of a polygon is the angle between any side extended and the adjacent side not extended. It is a positive number if the angle lies outside the polygon, and it is negative if it lies inside the polygon.* This definition is consistent with that used for convex polygons and furthermore gives the same sum for all polygons.

This article has shown that the sum of the interior as well as of the exterior angles of any polygon does not depend upon whether it is concave or convex.

2 The empty set or a set containing exactly one point are convex since there do not exist two distinct points X and Y such that $\overline{XY}$ is not in the set. That is, in these cases, the definition is vacuously satisfied.

3 If the student means any inequality, then the answer is yes. Consider the following:

$$\{(x, y) | x^2 + 1 > x^2\}.$$

This inequality is satisfied by all points in the plane. If the student is referring to *linear inequalities* such as

$$\{(x, y) | ax + by > c \text{ or } ax + by < c\}$$

where either a or b is not zero, then the answer is no.

***Problem*** Find two more inequalities that are satisfied by all points in the plane.

4 Each constraint in a linear programming problem yields a convex set. The feasible region is the intersection of these convex sets and hence is convex.

***Problem*** Prove that the intersection of convex sets is again convex.

## 3.6 Logic

### Questions

1 You ask your class if the following statement is true or false. "If not a square, then not a rhombus." One student says "if not a square" implies it could be a rectangle and a rectangle is not a rhombus. Therefore the statement is true. Do you agree?

2 You are beginning the proof of the statement that the diagonals of a parallelogram bisect each other. A student asks, "Don't we have to show that the diagonals intersect in the interior of the parallelogram?" Is this student correct?

3 The problem given to the class is as follows.

*Prove* If 2 lines are intersected by a transversal, such that the two alternate interior angles are congruent, then the lines are parallel.

*Given* $\overleftrightarrow{DC}$ and $\overleftrightarrow{EA}$ intersected by transversal $\overleftrightarrow{AC}$ so that $\angle EAC \cong \angle ACR$.

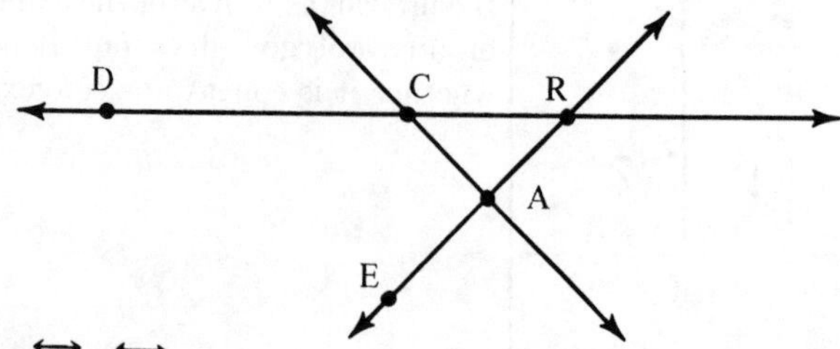


*Prove* $\overleftrightarrow{DC} \parallel \overleftrightarrow{EA}$.

A student offers the following proof. "Suppose $\overleftrightarrow{DC}$ is not parallel to $\overleftrightarrow{EA}$, then the two lines intersect at some point, say R. A triangle ACR is formed having an exterior angle EAC. According to a previous theorem, the measure of an exterior angle of a triangle is greater than either remote interior angle. Therefore

$$m\angle EAC > m\angle ACR,$$

but this contradicts the given that $\angle EAC \cong \angle ACR$. Therefore assumption that $\overleftrightarrow{DC}$ is not parallel to $\overleftrightarrow{EA}$ is false; therefore the two lines are parallel."

This proof could be used to prove any pair of lines are parallel. What is wrong with the theory? The above problem was given to a geometry class by a student teacher. The answer given by the student teacher was the following: "By assuming that willing something is so, if two lines are not parallel, then alternate interior angles could never be formed that were congruent, and the balance of the proof is fallcious." How would you react to this problem and the statement offered by this student teacher?

4 In your tenth grade class, you are now going over the truth table for a conditional statement. The students are having trouble understanding why the last two rows of the conditional are marked true (i.e., antecedent false, consequent true, and antecedent false, consequent false). How would you help them?

5 A student asks, "While studying these truth tables, we have assumed that a statement is either true or false. Is there a mathematical study of truth tables in which statements could be true, false, and possibly true?" How do you respond?

6 A student asks you if there is any connection between the biconditional and the idea of equivalent expressions that he learned in Algebra I. What is your response?

7 During a lesson on indirect proofs, a student says, "Why do we have to assume the opposite of the conclusion? Why can't we assume the opposite of the given and try to contradict the conclusion?" What is your reply?

8 You ask a student to give you the negation of "All men are good." The student says that the negation would be "All men are not good." Is the student right?

9 "If we can always use direct proof, why do we need to know about indirect proofs?" How would you answer the student?

10 A student asks, "How do we know that we are talking about straight lines when there has been no indication that a line must be straight. Couldn't it be curved?" What is your response?

11 You ask your students to put the following statement into "if-then" form. "Two planes intersect provided that they are not parallel." One student writes, "If two planes intersect, then they are not parallel." Is the student right?

12 A student makes the remark that it seems that a conditional statement and its con-

verse are related to the symmetric property of a binary relation. How do you react to this remark?

13 A student makes the following statement. "A sphere is the three-dimensional figure analogous to a circle, a two dimensional figure. Since at a point on a circle there is exactly one line tangent to the circle, there is exactly one line tangent to a sphere at a point on the sphere." Is this student correct?

14 A student asks, "Our book says that the geometry written by Euclid had defects in it, but the book does not say what they were. Could you tell me?" How would you reply?

15 A student says that the number of points on a line segment is finite. How can you convince the student that this is not correct?

16 "Is it ever possible in mathematics to prove a statement or theorem by proving a special case of the statement or theorem?" How do you respond to this student?

17 One of the problems in your textbook is stated as follows: "What is the negation of the statement 'Some men are not wise'?" A student replies that the negation would be "Some men are wise." Do you agree?

18 You have asked your class to write a false conditional so that its converse is also false. A student says that it is impossible to find one. How do you respond?

## Answers

1 No, the statement is not true. Perhaps, the best way of showing this fact is to look at the contrapositive of the given statement. Thus, suppose we look at the statement, "If a rhombus, then a square." This statement is certainly not true. Since this statement is equivalent to the given statement, then we can conclude that the given statement is not true.

*Problem* Where is the student having difficulty in this question?

2 Logically speaking, the student is correct. Usually, however, in a high school course in geometry, we assume that the diagonals do intersect in the interior of the parallelogram.

3 The proof is valid and the student teacher is incorrect. In fact, the argument given by the student teacher, i.e., if two lines were not parallel, then alternate interior angles could never be formed that were congruent, is the basis for arriving at the contradiction in this indirect proof.

4 One might make up an example which would help students see why it is reasonable that the last two rows of a conditional are marked true. Two examples are given below which have been found to be effective in accomplishing this aim.

*Example 1* Suppose a teacher in a certain algebra course makes the following statement. "If a student in this course gets A's on all four exams, then that student will get an A in the course."

*Case 3* Suppose a student in this course does not get A's on all four exams but still the student gets an A in the course. Has the teacher broken his promise to the students in this course? No, the teacher has not; he has still kept his word.

*Case 4* Suppose a student in this course does not get A's on all four exams and does not get an A in the course. The teacher still has not broken his promise; i.e., the original statement is still true.

*Example 2* Suppose that a man who lives in Maryland makes the following statement. If I live in Delaware, then I live in the United States. The hypothesis is false and the conclusion is true but still his statement is true. Suppose that a woman who lives in France makes the same statement as above. In this case, both the hypothesis and the conclusion are false, but the statement is still true.

5 Yes, there is. Kramer gives the following account. In 1921 two research papers appeared, one by the Polish mathematician Jan Lukasiewicz on three-valued logic, and one by the American Emil Post on general n-valued systems. Table 3.6.1 is an example of a three-valued truth table.*

*TABLE 3.6.1*

| $p$ | $q$ | $p \to q$ |
|---|---|---|
| T | T | T |
| T | M | M |
| T | F | F |
| M | T | T |
| M | M | M |
| M | F | M |
| F | T | T |
| F | M | T |
| F | F | T |

T True
F False
M Possible (not known to be true or false)

*Kramer, Edna. *The Nature and Growth of Modern Mathematics.* Hawthorn Books, Inc. New York, 1970, pp. 132-33.

6 Yes, the ideas are related, and they are really isomorphic systems.

Let p be a statement, and let P be an arbitrary set. Then under the correspondence shown below, set theory and logical analysis have essentially the same underlying structure.

(1) $p \to P$ (2) $\wedge \to \cap$
(3) $\vee \to \cup$ (4) $\sim \to$ Complement ($\overline{\ }$)

Thus, properties such as De Morgan's laws in set theory have a corresponding property in logical analysis.

(a) $\overline{A \cap B} = \overline{A} \cup \overline{B}$ $\quad \sim(p \wedge q) \leftrightarrow \sim p \vee \sim q$

(b) $\overline{A \cup B} = \overline{A} \cap \overline{B}$ $\quad \sim(p \vee q) \leftrightarrow \sim p \wedge \sim q$

One can go further in this respect as shown in Tables 3.6.2 and 3.6.3.

*TABLE 3.6.2*

| $p$ | $q$ | $p \wedge q$ |
|---|---|---|
| T | T | T |
| T | F | F |
| F | T | F |
| F | F | F |

*TABLE 3.6.3*

| $X \in P$ | $x \in Q$ | $x \in P \cap Q$ |
|---|---|---|
| T | T | T |
| T | F | F |
| F | T | F |
| F | F | F |

If p and q are closed statements, then p is equivalent to q ($p \rightarrow q$) whenever p and q have the same truth value. That is, when p is true, then q is true; when p is false, q is false. If $P_x$ and $Q_x$ are open statements, then $P_x$ is equivalent to $Q_x$ ($P_x \leftrightarrow Q_x$). Both open statements have the same solution set. That is, when $P_x$ is true then $Q_x$ is true; when $P_x$ is false, $Q_x$ is false.

7 One form of an indirect proof is to prove the contrapositive of a given conditional. Thus if one is trying to prove the conditional $p \rightarrow q$, then one might prove the contrapositive $\sim q \rightarrow \sim p$. Since a conditional and its contrapositive are logically equivalent, then the original conditional is valid. The student seems to be saying that one should also be able to prove a conditional $p \rightarrow q$, by proving the statement $\sim p \rightarrow \sim q$ to be true. However, the problem here is that these two statements are not logically equivalent. Thus proving the statement $\sim p \rightarrow \sim q$ to be a valid statement does not determine the validity of the statement $p \rightarrow q$.

In a more general sense, the idea of an indirect proof is as follows. One assumes that the given (p) is true and that the negation of q is true. The idea is then to try to show that these two statements lead to a false statement; i.e., $(p \wedge \sim q) \rightarrow s$ where s is a statement that is known to be false. Frequently s takes the form $r \wedge \sim r$, which is definitely a false statement. Since the reasoning was correct (the conditional is true), then by looking at a truth table for a conditional one finds that this is the case only if $p \wedge \sim q$ is false. But p is assumed to be true and thus $\sim q$ must be false, which in turn implies that q is true. Since p is assumed true and q is true, then $p \rightarrow q$ must be true.

What happens if we try to prove a conditional, $p \rightarrow q$ using the student's method. In this case, we would try to show that $\sim p \wedge q \rightarrow t$ where t is a false statement. As in the previous case, $\sim p \wedge q$ must be false; since we know that p is true, therefore $\sim p$ is false. However, this is as far as we can go because in this case q could be either true or false; i.e., we cannot deduce absolutely that q is true. Thus the method suggested by the student will not allow us to prove a conditional, $p \rightarrow q$ indirectly.

8 This is a tricky question to answer because we really do not know what the student

meant. The best advice that we can offer is to try to find out exactly what meaning the student is attaching to his statement. If you find out that the student is really saying that not all men are good, then the student is correct. However, if the student meant *all* (or every single) men are not good, then the student is incorrect. Thus, your evaluation of the student's response depends upon what the student emphasized in the response. You should certainly try to lead the student to a better interpretation, which has no ambiguous meaning attached to it such as "There is at least one man who is not good."

9 Although one might always be able to use the method of direct proof, there are some theorems that are extremely difficult to prove using this method, whereas an indirect proof could be quite simple.

10 No, the lines could not be curved. The postulates and theorems rule out the possibility of curved lines. For example, it is usually given as a postulate that through any two different points there is exactly one line. It is usually proved as a theorem that if two lines intersect, they intersect in exactly one point,These two statements, along with others,obviously rule out the possibility that "lines" are curved.

11 No, the student is incorrect. The correct form is: if two planes are not parallel, then they intersect.

12 The student is correct in seeing an analogy in this case, but caution should be exercised. The symmetric property of a binary relation is stated as follows:

> If R is a binary relation, then R is symmetric if $a\ R\ b \Rightarrow b\ R\ a$; i.e., if $a$ is in the relation R to $b$, then $b$ is in the relation R to $a$.

However, it does not have to be the case that *if p, then q* is true the converse, *if q, then p*, is true. Thus although the two situations are similar, they are not totally related.

13 No, the student is incorrect. There are an infinite number of lines tangent to a sphere at a point on the sphere. A correct statement can be made which shows the analogy. That is, at a point on a sphere there is exactly one plane tangent to the sphere.

14 (1) Euclid did not recognize the need for undefined terms. Instead of beginning with a few undefined terms whose meanings come from the axioms, he tried to define every term. Thus he defined line as follows: "A line is breadthless length." But what does breadthless mean?

(2) Euclid did not include axioms of incidence, order and congruence. Euclid assumed and used many properties that were not included in his list of axioms and could not be derived from them. For example, Euclid says nothing about the relation of betweenness and thus we do not know what the following statement means: point y is between points x and z. Also Euclid did not spell out incidence axioms, which express basic relationships of points, lines and

planes. For example:

(a) All lines and planes are sets of points.

(b) If two distinct planes intersect, then their intersection is a line.

Euclid made the concept of congruence depend on the concept of superposition, i.e., that a triangle can be picked up and put down in another place with all the properties remaining invariant. But no statement of this was made.

(3) Euclid's axioms say nothing about continuity yet he assumed that a line segment intersected a triangle in a point as shown in Figure 3.6.1. However, this idea depends upon continuity.

*FIGURE 3.6.1*

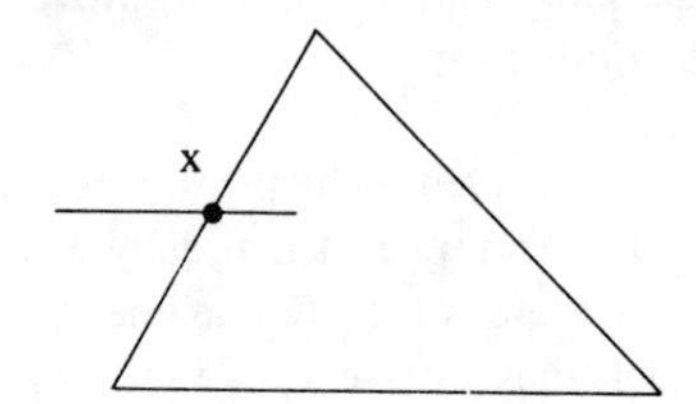


Euclid also assumed that a line is infinite in extent, but no axiom states this. He does state that "A finite straight line can be produced continuously in a straight line," but this does not imply that the line is infinite in extent. It implies only that it is boundless.

(4) Euclid assumed that if a straight line enters a triangle at a vertex, it must intersect the opposite side. An axiom stating this needed to be given.

15 One way of doing this is as follows. Suppose that segment $\overline{AB}$ contains a finite number of points. Without loss of generality, we can consider a coordinate system set up on line $\overleftrightarrow{AB}$ such that the coordinate of A is zero and the coordinate of B is one. It is a well-known fact that there exists a one-to-one correspondence between the set of real numbers and the set of points on $\overleftrightarrow{AB}$. In particular, there is a one-to-one correspondence between the points of $\overline{AB}$ and the set of all real numbers between 0 and 1. Thus if $\overline{AB}$ contains only a finite number of points, then this would imply that there is only a finite number of real numbers between 0 and 1. However, this last statement contradicts the well-known fact that there is an infinite number of real numbers between 0 and 1. Thus, the number of points on a line segment is infinite.

16 Generally, the answer is no. In fact, at this level, proving a special case and then assuming that the general case has been proved is one of the common mistakes made by students, e.g., using coordinates that depict a square when proving a theorem on rectangles. However, in some cases, when one can show that a general statement is equivalent to a special case, then one can prove the special case and thus assert that the general case is valid. Polya makes this point in his book, *Mathematics and Plausible Reasoning*, Vol. 1. He shows that the equation $a^2 = b^2 + c^2$ is equivalent to the equation $\lambda a^2 = \lambda b^2 + \lambda c^2$. He states, "The general theorem is equivalent not only to the special case ($a^2 = b^2 + c^2$), but to any other special case. Therefore,

if any such special case should turn out to be obvious, the general case would be demonstrated." He goes on to say, "It shows also the fact, so usual in mathematics and still so surprising to the beginner...that the general case can be logically equivalent to a special case."*

17 No. The statement made by the student, "Some men are wise," implies the possibility that there are some men who are not wise, i.e., at least one man who is not wise, which is not the negation of the given statement. A correct negation would be "Every man is wise."

18 One way of looking at the question is to look at the truth table for a conditional as given in Table 3.6.4.

*TABLE 3.6.4*

| $p$ | $q$ | $p \to q$ |
|---|---|---|
| T | T | T |
| T | F | F |
| F | T | T |
| F | F | T |

If a conditional is false, then we are looking at the second row in the truth table; i.e., if $p \to q$ is false, then p is true and q is false. Therefore, the converse would be $q \to p$ where q is false and p is true. But in this case, $q \to p$ is a true statement. Thus the student is correct.

*Gyorgy Polya, *Mathematics and Plausible Reasoning*, Vol. 1, Princeton University Press, Princeton, N.J., 1954, p. 17.

# 3.7 Midpoints (Segments and Rays)

## Questions

1 A student asks you why lines or rays don't have bisectors as segments do. In response, you tell the student that rays or lines do not have midpoints. The student then looks at you puzzled and asks you why rays or lines do not have midpoints? How do you respond?

2 A student is trying to find the midpoint of $\overline{AB}$. The coordinate of A is x, and the coordinate of B is y with $y > x$. The student reasons as follows: "The coordinate of the midpoint m of $\overline{AB}$ should be ½ of the distance between A and B. Therefore the coordinate y of the midpoint m is $\frac{1}{2}(y - x)$." Is the student correct?

## Answers

1 A midpoint of a segment $\overline{AB}$ is defined to be that point X such that AX = BX. The question then is whether this idea can be extended to include rays and lines. However, one can probably see that if such a point X existed for either a ray or a line, then there exist infinitely many points that could be classified as midpoints. Since a unique answer is not available, it is customary to say that rays and lines do not have midpoints.

2 No, the student is not correct. The coordinate z of the midpoint m is

$$x + \tfrac{1}{2}(y - x) = x + \tfrac{1}{2}y - \tfrac{1}{2}x = \tfrac{1}{2}x + \tfrac{1}{2}y = \tfrac{1}{2}(x + y)$$

# 3.8 *n*-Dimensional Space, $n \geq 3$

## Questions

1 "We have been talking about simple closed curves in two-dimensional space. What would be the analogue in three dimensions of this idea?" How do you respond to this student?

2 In your geometry class, the following question arises. "In mathematics is there a fourth, fifth, and *n*th dimensional space? If so, is there any use for such spaces?"

3 A student asks, "If a regular polyhedron is made up of regular polygons and there are an infinite number of regular polygons, then why are there only five regular polyhedrons?" How do you respond?

## Answers

1 In three dimensions the analogue would be any closed surface which does not cross itself. The ideas are connected by Jordan's theorem. In two dimensions the theorem states that "any closed curve in the plane, which does not cross itself, divides the plane into an inside and an outside." In three dimensions, the theorem states that "any closed surface, which does not cross itself, divides space into an inside and an outside." For more information on this idea consult:

Edward Kasner and James Newman, *Mathematics and the Imagination,* Simon and Schuster, New York, 1963, pp. 265-98.

2 Mathematically, there are spaces of dimension $n$ where $n$ is any positive integer. This is a major concern in the study of vector spaces. Indeed, mathematicians now study infinite dimensional spaces. Such studies have, for example, led to a greater knowledge of polynomial approximations of rather complicated functions.

Einstein suggested that in order to study the universe in which we live one must regard space as four dimensional. In addition to three position coordinates, one should also add a fourth dimension of time. In many engineering applications of mathematics, one deals with several variables, say position in space (three variables), time, temperature, velocity, etc., so that one is frequently working in a space of dimension greater than three.

3 The reason that there are only five regular polyhedra is based upon the fact that the sum of the measures of the angles at each vertex of a geometric solid must be less than 360°. (Why?) Remember that a regular polyhedron is a geometric solid, all of whose faces are congruent regular polygons in which the same number of polygons meet at each vertex. Now, at least three faces must meet at each vertex of a polyhedron. (Why?) Thus since each angle of a regular hexagon has a measure of 120° there is no regular polyhedron whose faces are regular hexagons. Since each angle of a regular pentagon has a measure of 108°, then there are no regular polyhedron in which four pentagons meet at each vertex. Table 3.7.1. gives each regular polyhedron with its corresponding number of regular polygons at each vertex.

*TABLE 3.7.1*

| *Polyhedron* | *Number of Regular Polygons at Each Vertex* | *Reason* |
|---|---|---|
| Tetrahedron | 3 equilateral triangles | $3 \cdot 60° < 360°$ |
| Cube | 3 squares | $3 \cdot 90° < 360°$ |
| Octahedron | 4 equilateral triangles | $4 \cdot 60° < 360°$ |
| Dodecahedron | 3 pentagons | $3 \cdot 108° < 360°$ |
| Icosahedron | 5 equilateral triangles | $5 \cdot 60° < 360°$ |

## 3.9 Non-Euclidean Geometry

### Questions

1 A student asks, "In Euclidean geometry, why is a distinction sometimes made between the *real* Euclidean plane and the *complex* Euclidean plane?" How do you reply?

2 A student asks, "Do we live in a Euclidean or non-Euclidean world?" How do you respond?

3 "What distinguished Euclidean geometry from non-Euclidean geometry?" How would you answer this student?

## Answers

1 The *complex* Euclidean Plane, though not studied at the secondary school level, is defined to be the set of all points $(z_1, z_2)$ where $z_1$ and $z_2$ are complex numbers. Thus the *complex* Euclidean Plane is an extension of the *real* Euclidean Plane, which is defined to be the set of all points (x, y) where x and y are real numbers.

This extension to complex values makes it possible to consider theorems in a more general setting and avoid many of the constraints that are imposed by a restriction to real values. For more information consult:

(1) Felix Klein, *Elementary Mathematics From An Advanced Standpoint — Geometry*, Dover Publications, New York, 1939, pp. 117-29.

(2) Rafael Artzy, *Linear Geometry*, Addison-Wesley Publishing Co., Reading, Ma., 1965.

2 If we assume that something is uniformly distributed over a surface, we can determine whether the surface is Euclidean (flat) or non-Euclidean (curved) by measuring numbers of a given thing at varying distances from a given point on the surface. This is one of the methods that astronomers have used to try to determine whether space is Euclidean or non-Euclidean.

Radio astronomers at Ohio State University have discovered that the farther out in space one looks, the more the number of observed radio sources falls short of what one would expect if space were Euclidean. If one assumes that radio sources are uniformly distributed in space, an implication that can be drawn is that space is non-Euclidean, specifically, Riemannian as Einstein has formulated.*

However, over relatively small regions, Euclidean and non-Euclidean methods give essentially the same results. Whether space is Euclidean or non-Euclidean is usually not the domain of mathematicians but is best for physicists and astronomers to ponder.

*Adapted from Harold Jacobs, *A Teacher's Guide to Geometry*, W.H. Freeman and Co., San Francisco, 1974, p. 96.

3 The main point that distinguishes these two is the parallel postulate. In Euclidean geometry, it is assumed that through a point P not on line $\overleftrightarrow{AB}$, there is only one line through P parallel to $\overleftrightarrow{AB}$. In non-Euclidean geometry modifications of this postulate plus other modifications are made. Thus in elliptic or Riemannian geometry, it is assumed that two straight lines always intersect each other; i.e., through a point P not on $\overleftrightarrow{AB}$, there are no lines parallel to $\overleftrightarrow{AB}$. In hyperbolic geometry, it is assumed that through a point P not on $\overleftrightarrow{AB}$ there exists more than one line that does not in-

tersect $\overleftrightarrow{AB}$. For more details on this question the following references are recommended:

(1) E.T. Bell, *Men of Mathematics*, Simon and Schuster, New York, 1937.

(2) Courant and Robbins, *What is Mathematics?* Oxford University Press, New York, 1941.

(3) Howard Eves, *A Survey of Geometry*, Vol. 1, Allyn and Bacon, Boston, 1963.

(4) Edwin Moise, *Elementary Geometry from an Advanced Standpoint*, Addison Wesley Publishing Co., Reading, Ma., 1963.

(5) Harold E. Wolfe, *Non-Euclidean Geometry*, Holt, Rinehart and Winston, New York, 1945.

(6) *Topics for Mathematics Clubs*, N.C.T.M., Reston, Va., 1973, Chap. 8.

## 3.10 Parallelism

### Questions

1 A student asks you if it is possible for two lines to be parallel and for the lines to be contained in intersecting planes. What is your response?

2 A student asks you if parallel lines meet at infinity. What is your response?

3

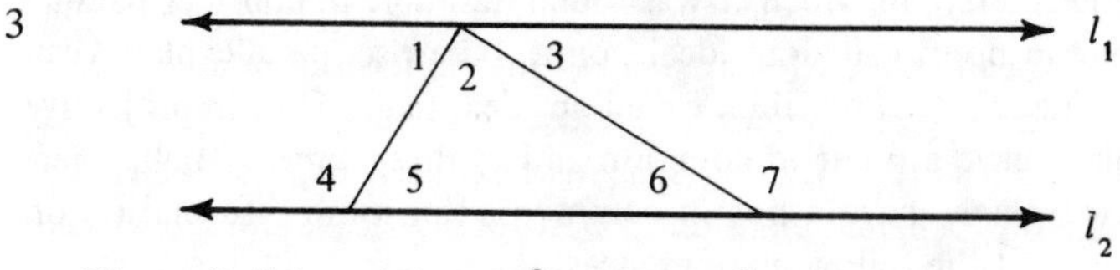


*Given* $l_1 \parallel l_2$, $m\angle 7 = 150°$, $m\angle 5 = 70°$. Find $m\angle 1$.

A student gives the following answer. "$m\angle 1 = 150°$ because it is an angle alternate interior to $\angle 7$." Is the student correct?

4 One student says that segments which have no points in common are either parallel of skew segments. Do you agree?

5 A student asks, "In the construction of a brick school building, a bricklayer uses a plumb bob to be sure the corners are all vertical. Are the corners (edges) then parallel?" How do you respond?

6 "In our text it says to extend line $\overleftrightarrow{AB}$... How can you extend a line when a line is infinite in nature? I don't understand." How would you help this student?

7 A student is working on the following problem. "Name segments that can be deduced to be parallel in the figure if $m\angle 8 = m\angle 4$."

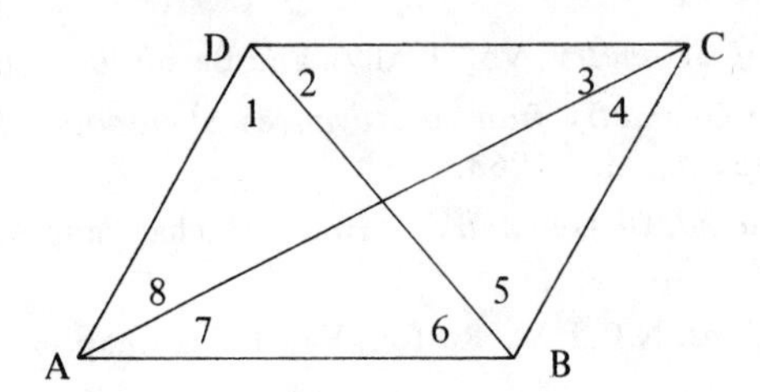


One student says that segments $\overline{AD}$ and $\overline{BC}$ are parallel and segments $\overline{AB}$ and $\overline{CD}$ are parallel. Is the student correct?

## Answers

1 Yes, it is possible. Let $l_1$ and $l_2$ be two parallel lines contained in plane $P_1$. Let $P_2$ be a plane that intersects $P_1$ where the line of intersection is $l_1$. Then $P_1$ and $P_2$ are intersecting planes; $l_1$ is contained in $P_2$; and $l_2$ is contained in $P_1$.

***Problem*** Construct an example of the above different from the one given.

2 In ordinary Euclidean geometry parallel lines are not considered as "meeting at infinity." However, there are many different kinds of geometry and there is a geometry, projective geometry, in which it was found desirable to consider parallel lines as having a common point called an ideal point. Likewise, parallel planes can be considered as meeting at a distinct line, called an ideal line. Thus in projective geometry, any two lines have a point in common and in three-dimensional projective geometry any two planes have a line in common. For more information on projective geometry consult the following references.

(1) *Topics for Mathematics Clubs*, N.C.T.M., Reston, Va., Chap. 2, 1973.

(2) Morris Kline, "Projective Geometry," *Mathematics in the Modern World, Readings from Scientific American*, W.H. Freeman and Co., San Francisco, 1968.

(3) David Gans, *Transformations and Geometries*, Appleton-Century-Crofts, New York, 1968.

3 No, the student is not correct. If we break the diagram apart and show the two transversals separately, one can see that $\angle 1$ is an alternate interior angle to $\angle 5$. Thus $m\angle 1 = 70°$. Angle 1 is not an alternate interior angle to angle 7.

4 No. Consider Figure 3.10.1. $\overline{RS}$ is neither parallel to, nor skew to, $\overline{XY}$.

*FIGURE 3.10.1*

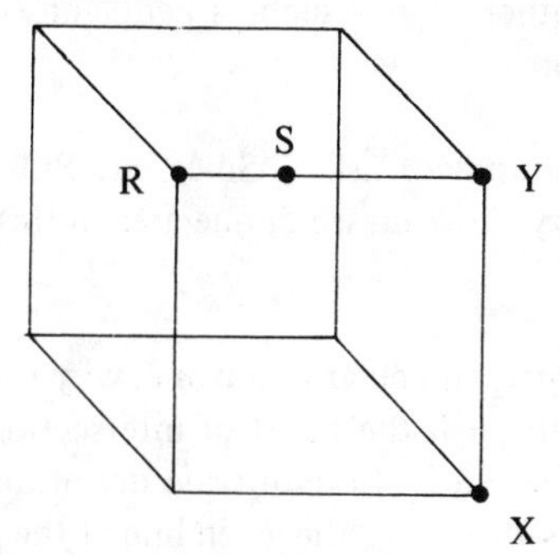


5 For the practical purposes of the bricklayer, the answer is *yes*. However, from a theoretical point of view, the answer is *no*. For suppose Figure 3.10.2 represents a "side view" of the earth. The plumb bob, if properly affixed, points to the center of the earth, and on a large scale, one would obtain the nonparallel results indicated.

*FIGURE 3.10.2*

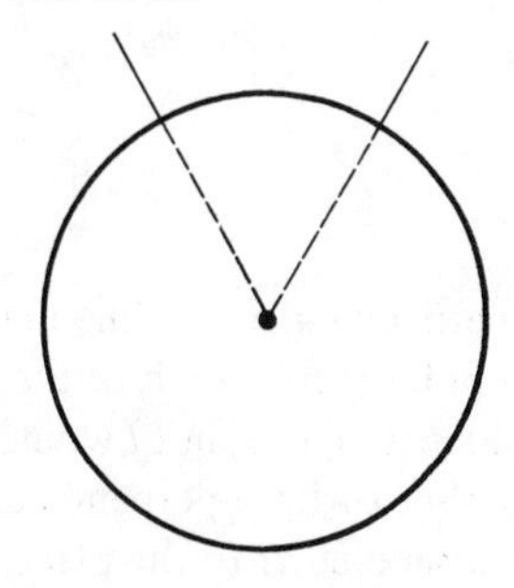

6 The statement made by the text has to be read in context. The main difficulty here is one between a geometric entity (a line) and its representation. What the statement in the book means is to extend our representation for a line (the figure). The student is correct in saying that we cannot extend a line. It is the representation, not the line itself, that is extended.

7 No, the student is not correct. Only segments $\overline{AD}$ and $\overline{BC}$ can be deduced to be parallel. In order to say that $\overline{AB}$ and $\overline{CD}$ are parallel, one would have to know that $m\angle 7 = m\angle 3$. But one cannot make this assertion without additional information.

## 3.11 Perpendicularity and the Pythagorean Theorem

### Questions

1 A student asks, "If a line, *l*, is perpendicular to a line in the plane, is *l* perpendicular to the plane?" How do you answer?

2 "When studying the Pythagorean Theorem, we learned that the square constructed on the hypotenuse equals the sum of the squares constructed on the other two sides. Do similar relationships hold for other figures such as pentagons or hexagons?" What is your response to this student.

3 While studying non-Euclidean geometry, a student asks, "Since the Pythagorean Theorem is not true in hyperbolic geometry, how do we define area in this geometry?" How do you respond?

4 A student asks, "In the definition of a line perpendicular to a plane, why is it necessary that every line in the plane passing through the point of intersection of the given line and plane be perpendicular to the line? Couldn't the definition be restated with just one line in the plane perpendicular to the given line at the point of intersection, or two lines?" How would you answer this student?

## Answers

1 No, it is not. Consider a line *l* that is vertical and perpendicular to a line m at point r in plane P. If *l* is now rotated about m through point r, then each of the infinite positions thus generated will still have *l* perpendicular to m at r, but *l* would not be perpendicular to P. Thus there are an infinite number of lines perpendicular to a given line in a plane but only one of these lines is perpendicular to the plane.

2 Yes, it does. Consider three similar figures constructed on the sides of a right triangle as shown in Figure 3.11.1.

*FIGURE 3.11.1*

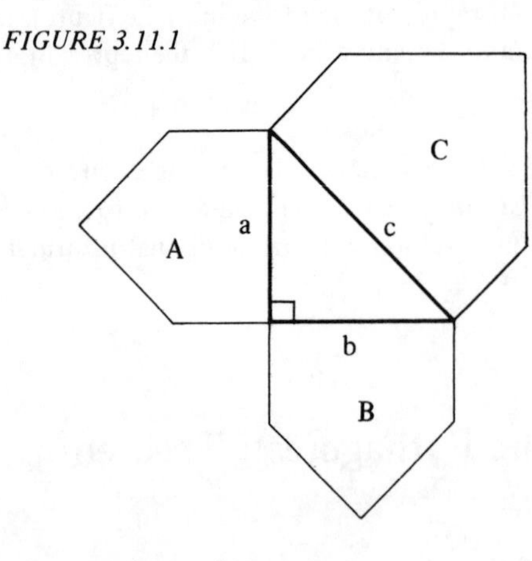


$$\frac{\text{Area A}}{\text{Area C}} = \frac{a^2}{c^2} \qquad \text{Why?}$$

$$\frac{\text{Area B}}{\text{Area C}} = \frac{b^2}{c^2} \qquad \text{Why?}$$

$$\frac{\text{Area A}}{\text{Area C}} + \frac{\text{Area B}}{\text{Area C}} = \frac{a^2}{c^2} + \frac{b^2}{c^2} = \frac{a^2 + b^2}{c^2}.$$

But $c^2 = a^2 + b^2$, thus

$$\frac{\text{Area A} + \text{Area B}}{\text{Area C}} = \frac{c^2}{c^2} = 1;$$

i.e., Area A + Area B = Area C.

3 Any definition of area should satisfy the following properties.

(1) To each region there is assigned a unique real number called the area of the region.

(2) Regions that are congruent must have the same area.

(3) If a region is split into a finite number of nonoverlapping parts, then the area of the region is the sum of the areas of these parts.

Any function that assigns to each geometric figure a number that satisfies the above three properties is called an area function.

In Hyperbolic Geometry, an area function is defined as the amount by which the sum of the angles of a triangle differs from 180°. This area function is usually called the defect of the triangle.

4 For the case of one line refer back to Answer 1 in this section. In the case of two lines, the story is quite different. In fact in some texts, the following statement is given as a theorem.

If a line is perpendicular to each of two intersecting lines at their point of intersection, then it is perpendicular to the plane that contains them.

***Problem*** Prove the statement given above.

# 3.12 Polygons

## Questions

1 A student asks, "Since an equilateral triangle must also be an equiangular triangle, then is an equilateral hexagon also equiangular?" What is your answer?

2 "If two polygons have their corresponding angles equal, are their sides proportional?" How do you answer this student?

3 A student asks, "What is the relationship between parallelograms, trapezoids, rectangles, squares, and rhombuses." What is your response?

4 A geometry book defines a trapezoid as a quadrilateral with *exactly* two sides parallel. A student asks why it can't be defined as a quadrilateral with at least two sides parallel? What is your response?

5 A student makes the following statement. "It looks like if you join the midpoints of the sides of a quadrilateral, the resulting inscribed figure is always a parallelogram." Is this a true statement? What is your response?

6 A student asks, "In a triangle, an exterior angle is greater than each of its remote interior angles. Is this statement true if we replace triangle with quadrilateral, pentagon, hexagon, etc.?" How would you respond?

7 "Is the area of a geometric (two-dimensional) figure always smaller than its perimeter? Likewise, is the volume of a solid always smaller than its surface area?" How do you respond to this student?

8 You are discussing the idea of exterior angles of polygons. One student asks you why the following is not an exterior angle of a polygon since it is *outside* of the polygon. How would you answer?

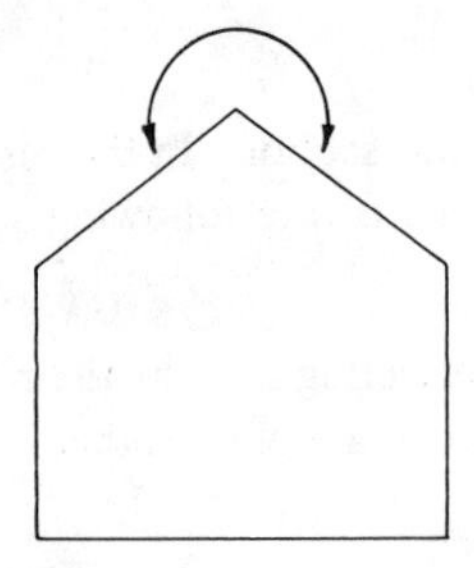

## Answers

1 Not necessarily; consider Figure 3.12.1. The figure is an equilateral hexagon, but it is not equiangular.

*FIGURE 3.12.1*

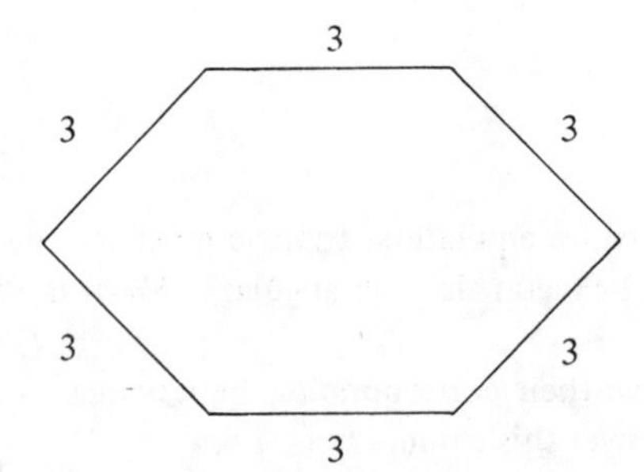

2 Not necessarily; consider Figure 3.12.2. These polygons have corresponding angles congruent, but their sides are not proportional.

FIGURE 3.12.2

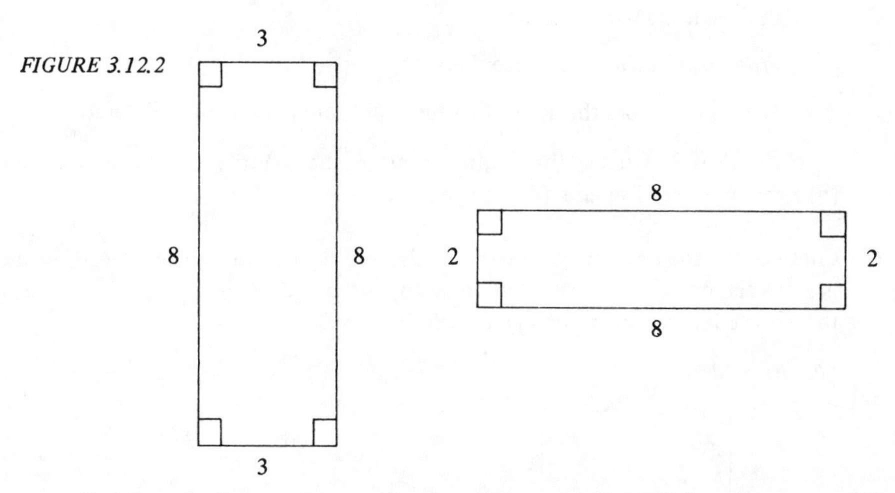


***Problem 1*** Construct two other examples using figures other than rectangles.

***Problem 2*** If the lengths of the corresponding sides of two polygons are proportional, are the polygons similar?

3 If a trapezoid is defined as a quadrilateral with exactly two sides parallel, then the relationship is as shown in Figure 3.12.3 where a name under another name means that the second type of named figure is a special case of the first figure named.

FIGURE 3.12.3

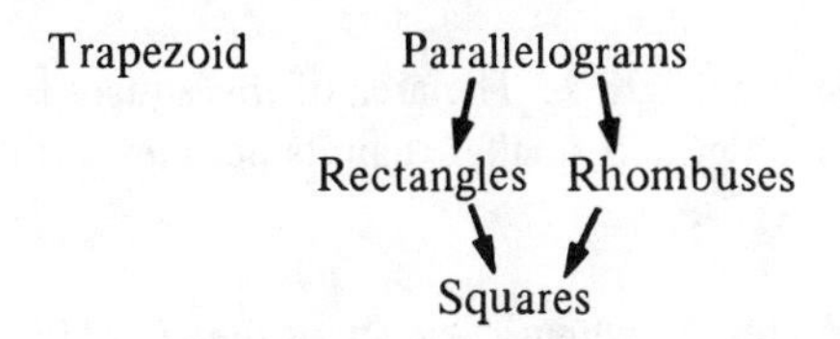


***Problem*** How would Figure 3.12.3 be changed if a trapezoid is defined to be a quadrilateral with two sides parallel?

4 In some geometry texts, a trapezoid is defined either as a quadrilateral with at least two sides parallel or as a quadrilateral with exactly two sides parallel. According to the first definition a parallelogram is then a special type of trapezoid. Which definition is used (exactly two sides parallel or at least two sides parallel) seems to be arbitrary and is usually left up to the discretion of the authors of a particular geometry text.

For more information on this question consult the following articles.

(1) David Dye, "What is a Trapezoid," *The Mathematics Teacher*, November 1967, pp. 727-28.

(2) Don Ryoti, "What is an Isosceles Trapezoid," *The Mathematics Teacher*, November 1967, pp. 729-30.

5 Yes, the statement is true and is known as Varignon's Theorem:

If the midpoints of consecutive sides of a quadrilateral ABCD are connected, the result is a parallelogram.

***Problem 1*** Prove the above theorem.

***Problem 2*** Does the above theorem hold for concave quadrilaterals?

***Problem 3*** What is the resulting figure if the quadrilateral is: (a) a rectangle, (b) a rhombus, (c) a square, (d) a trapezoid?

6 Consider the square shown in Figure 3.12.4 and the exterior angles at A. The measure of each exterior angle at A is equal to, not greater than, the measure of each of the remote interior angles of square ABCD.

*FIGURE 3.12.4*

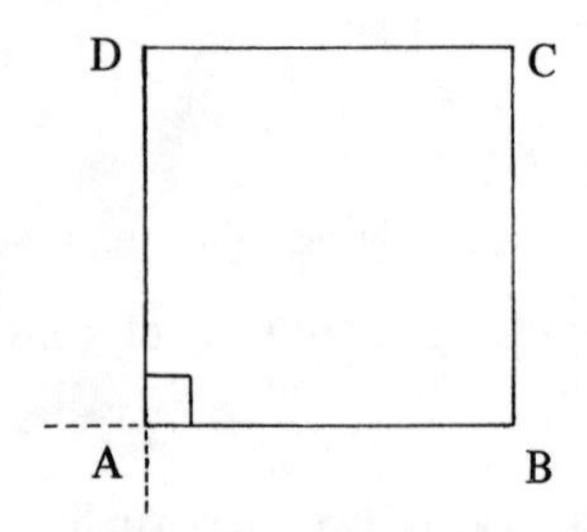


***Problem 1*** Construct two examples different from the one shown above.

***Problem 2*** Is the answer the same for all polygons of n sides where $n \geqslant 5$?

7 No. Consider a square whose side has length x. The area of this square is $x^2$ while its perimeter is 4x. If the area of this figure is smaller than its perimeter, then

$$x^2 < 4x,$$

whose solution set is $\{x | x < 4\}$. That is, when $x \geqslant 4$, then the area of a square is greater than its perimeter.

***Problem 1*** Give two other examples different from the one above that you might use to answer this question.

***Problem 2*** Answer the second part of the student's question.

8 If the angle referred to in this question is defined to be an exterior angle, then the measure of this angle is greater than 180°. However, in plane geometry, it is shown that the measure of an angle (with the usual definition of angle) is between 0° and 180°.

***Problem*** What is the usual definition of an exterior angle in plane geometry?

# 3.13 Similarity

## Questions

1 A student asks you if the SSS, SAS and AAA similarity theorems for triangles also hold for other polygons. What is your response?

2 You are comparing sides of similar triangles, and you ask your class to find EC as shown in the following figure.

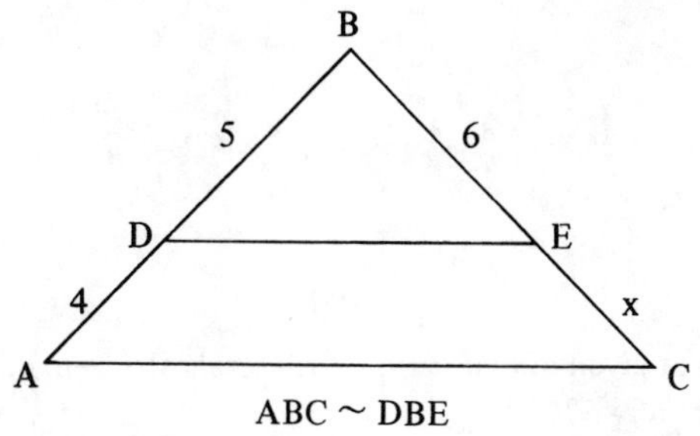


ABC ~ DBE

You show that $\frac{5}{9} = \frac{6}{6+x}$ or 5x = 24. A student shows his way:

$\frac{5}{4} = \frac{6}{x}$ or 5x = 24.

How do you handle this situation?

3 While studying a unit on similar triangles, a student asks, "Are there other uses for the projection of a segment on a line?" How do you respond?

4 A student asks you if similar polygons must also be regular polygons. How would you reply?

## Answers

1 If the polygons considered are quadrilaterals, then Figures 3.13.1 and 3.13.2 show that quadrilaterals must have both congruent angles and proportional sides to have similarity. In Figure 3.13.1, the corresponding rectangles have congruent angles, but the rectangles are not similar.

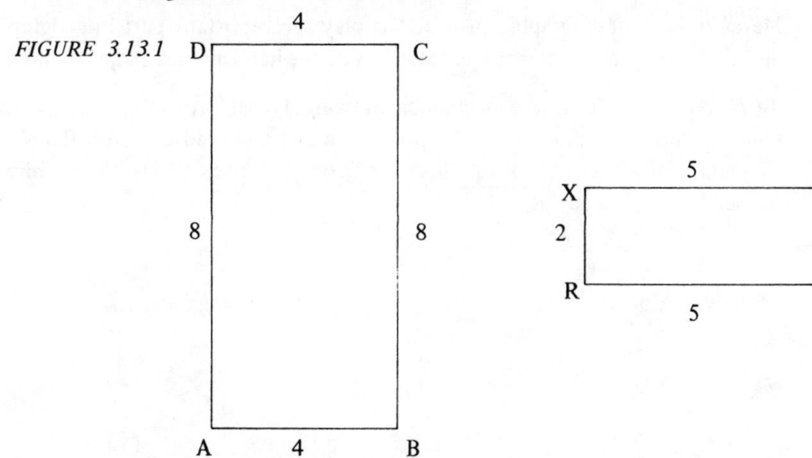


*FIGURE 3.13.1*

In Figure 3.13.2, the corresponding sides are proportional, but the figures are not similar.

*FIGURE 3.13.2*

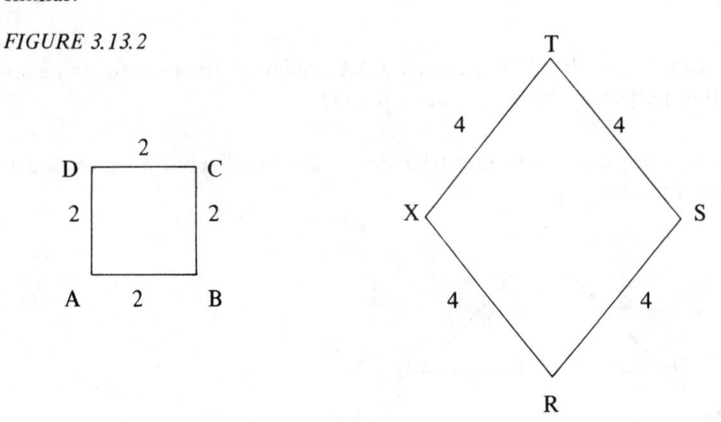


***Problem*** Does the answer given above for quadrilaterals also remain valid for other polygons of n sides where $n \geqslant 5$?

2 How you handle this situation depends on several factors, but ultimately comes down to what you are trying to accomplish in your teaching through this problem. It seems as if the teacher is trying to show how similar triangles can be used to solve a given problem. The student's method is correct but it is based on the following assertion.

> If a line is parallel to one side of a triangle and intersects the other two sides, it divides them proportionally.

If this assertion has been previously discussed, then the student is correct in using it. However, if this assertion has not been previously discussed, then there are several options open to you. Perhaps the best option is to ask the student to justify the way that the student solved the problem since this method would lead to the discovery and proof of the assertion stated above.

3 The idea of projections has many applications in the physical world. Some of these applications are:

(1) Mercator and stereographic projections play an important part in geography and astronomy as in atlas representations of the hemispheres and map making.

(2) In various industries where mechanical drawing is used projections are involved. One of these uses is to show the positioning of a curved cut on a flat sheet of metal. Another use is to show the geometric projections of dropping a perpendicular to a surface.

(3) Projective geometry is a branch of mathematics that arose primarily from the idea of projections. In his excellent book, Morris Kline shows how projections were used by Renaissance artists (such as Albert Durer) to achieve realism in their paintings.
Morris Kline, *Mathematics: A Cultural Approach*, Addison-Wesley Publishing Co., Reading, Ma., 1962, Chaps. 10 and 11.

4 No. Consider Figure 3.13.3. These figures are similar, but not regular.

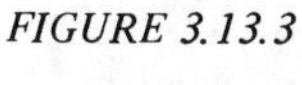
*FIGURE 3.13.3*

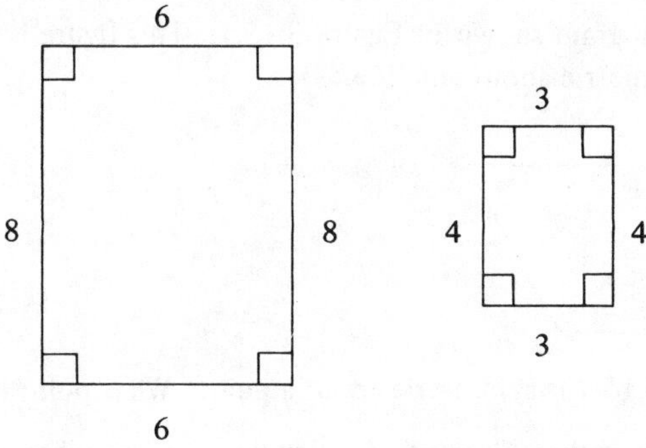


***Problem*** Construct two more examples different from the above. (Do not use rectangles.)

# 3.14 Skewness

## Question

1 A student says that any two lines that are in different planes are skew lines. Is this student correct?

## Answer

1 No, the student is not correct. Consider a plane such as represented by the floor of a room and a plane such as represented by the ceiling of the same room. Let a third plane, represented by the front wall of the room, intersect the given parallel planes. Each intersection will be a line, and even though these two lines are in different planes, they are not skew lines.

***Problem*** Modify the student's statement so that it is correct.

## 3.15 Symmetry

### Question

1 A student asks you if it is possible for a figure to be symmetric about a point but not symmetric about any line. What is your response?

### Answer

1 Yes. Consider the parallelogram shown in Figure 3.15.1. This figure is symmetric about a point but not symmetric about any line.

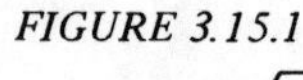

*FIGURE 3.15.1*

***Problem 1*** Figure 3.15.1 is symmetric about a point. What point is that?

***Problem 2*** Draw two other figures that are symmetric about a point but not symmetric about any line.

***Problem 3*** Is it possible for a figure to be symmetric about a line but not symmetric about any point?

***Problem 4*** Give a precise definition of each of the following: (a) symmetry with respect to a point, (b) symmetry with respect to a line.

## 3.16 Transformations and Functions

### Questions

1 While you are teaching geometry through a transformational approach, a student asks you if the composition $\alpha \circ \beta$ of two mappings $\alpha$ and $\beta$ can be one-to-one if neither $\alpha$ nor $\beta$ is a one-to-one mapping. How do you reply?

2 A student asks you if the composition $\alpha \circ \beta$ of two mappings $\alpha$ and $\beta$ can be an *onto* mapping if neither $\alpha$ nor $\beta$ is an onto mapping. How do you reply?

3 You have just shown that the set of isometrics (distance-preserving transformations) form a group. A student in your class makes the following comment. "A couple of weeks ago we showed that congruence of triangles is an equivalence relation. It

seems that what we have just shown is related to that idea by the following:

congruence ⟶ transformation
reflexive ⟶ identity isometry
symmetric ⟶ inverse isometry
transitive ⟶ product of isometries

Am I correct in my thinking?" How do you respond?

4 A student asks, "Since the *composition* of two translations is *always* a translation, why can a translation be *decomposed* into two transformations that are *not* translations?" How do you respond?

5 "If a glide reflection is defined to be the composition of a line reflection and a translation (or glide) in a direction parallel to the axis of reflection, what is the composition when the translation is *not* parallel to the axis of the reflection?" How do you reply to this student?

6 A student asks, "The composition of two rotations about the same point as center is always a rotation about that point. What is the composition of two rotations about different points as centers?" How do you respond?

7 "What are examples of transformations that leave lines or other figures fixed but do not leave all their points fixed?" How do you respond to this student?

8 A student asks, "If 'figure' is defined to be any set of points in the plane, (a) would mean that a set of *exactly two* points is a figure? (b) If so, what would be the meaning of saying one figure of two points is congruent to another figure of two points?" How do you respond?

9 A student asks, "If a square ABCD is folded on either of its diagonals, it is mapped onto itself. Does this occur for a rhombus which is not square?" What is your answer?

## Answers

1 This depends upon the definition of the composition of two functions that is being used. Two common definitions are given below.

(1) If $\alpha$ and $\beta$ are mappings, then $\alpha \circ \beta$ is the mapping given by $[\alpha \circ \beta](x) = \alpha(\beta(x))$ for all $x \in$ domain of $\beta$ for which $\beta(x) \in$ domain $\alpha$.

(2) If $\beta$ is a mapping with domain X and range Y and if $\alpha$ is a mapping with domain containing Y and range Z, then the composition of $\alpha$ and $\beta$, denoted by $\alpha \circ \beta$, is defined on the domain X and $[\alpha \circ \beta](x) = \alpha(\beta(x))$.

If Definition 1 is used, then the answer to the student's question is yes. Consider

$$\alpha(x) = \frac{x^2}{x^2 - 1} \qquad \text{for x a real number, } x \neq 1, -1,$$

and

$\beta(x) = 1$ for all $x \neq 2$
$0$ for $x = 2$.

Then according to Definition 1, $\alpha \circ \beta$ is the mapping that has domain consisting of $\{2\}$ and range consisting of $\{0\}$; i.e., it is one-to-one.

If Definition 2 is used, then the answer to the student's question is no. The domain of $\beta$ is the same as the domain of $\alpha \circ \beta$. Thus if $\beta(x_1) = \beta(x_2)$, $x_1 \neq x_2$, then $[\alpha \circ \beta](x_1) = (\alpha \circ \beta)(x_2)$.

***Problem 1*** Suppose that $\alpha$ is not one-to-one, but $\beta$ is one-to-one. Can $\alpha \circ \beta$ be one-to-one? Consider both definitions.

***Problem 2*** Suppose that $\alpha$ is one-to-one, but $\beta$ is not one-to-one. Can $\alpha \circ \beta$ be one-to-one? Consider both definitions.

2 Whether Definition 1 or 2 is used (see Answer 1 of this section), the answer is no. Since $\beta$ is not "onto," then $\alpha \circ \beta$ cannot be onto.

***Problem*** Suppose that $\alpha$ is not "onto," but $\beta$ is onto. Can $\alpha \circ \beta$ be onto?

3 Yes, the student is correct. The identity isometry (every figure is mapped onto itself) makes congruence reflexive. Since every isometry has an inverse, then congruence is symmetric and the transitive property of congruence depends on the idea that the product of two isometries is an isometry. Thus one can define two figures X and Y as congruent if there exists an isometry that maps X onto Y.

4 The student is assuming that a conditional is equivalent to its inverse; i.e., $p \to q$ and $\sim p \to \sim q$ are equivalent. Thus in this case, the student is really saying that if (1) below is valid then (2) is also valid.

(1) If r and t are translations, then $r \circ t$ is a translation.

(2) If r or t are not translations, then $r \circ t$ is not a translation.

Logically, however, a conditional and its inverse are not related in this manner. Thus even though (1) is true, (2) may or may not be true. In fact, in this particular case, it can be shown that if r and s are reflections over lines m and n where $m \parallel n$, then $r \circ s$ is a translation.

5 The composition, in either order, of a reflection and a translation when the translation is not parallel to the axis of the reflection is a reflection.

***Problem*** Prove the statement given above.

6 The composition of two rotations about different points as centers is either a rotation or a translation.

***Problem 1*** Find two rotations about different points that again form a rotation.

***Problem 2*** Find two rotations about different points that form a translation.

***Problem 3*** Prove the statement made in Answer 6.

7 There are numerous examples that could be given; a few are cited:

(a) Any reflection of the Euclidean plane across a line will leave the line fixed.

(b) One can prove that any isometry that leaves each of two points fixed must leave every point of their line fixed. Thus all one must do is locate two points that are fixed by an isometry, and the line determined by these two points is a fixed line under that isometry.

(c) Any rotation or reflection will leave a circle in the plane fixed.

(d) In the plane, a reflection over a line *l* followed by a translation parallel to the line *l* leaves the line *l* fixed. This transformation is called a glide reflection, and the fixed line is called the axis of the glide-reflection.

(e) Any reflection over a line g of a figure symmetric with g will leave the figure fixed.

***Problem 1*** Find two more examples not given above.

***Problem 2*** In (d), if the translation is performed first, and then the reflection, will *l* still be a fixed line?

8 According to the definition given, a set of two points would be a figure (albeit a trivial one). A figure F of two points is congruent to another figure G of two points if and only if one can be mapped onto the other by an isometry; i.e., there is a one-to-one correspondence between the points of F and G such that the distance between the two points of F equals the distance between their images in G.

9 Yes. One can show that a rhombus is reflection symmetric to the lines containing its diagonals. A geometric figure in the plane is reflection symmetric if and only if there is a reflecting line r such that the figure and its image over r coincide. The line r is called a line of symmetry of the figure.

## 3.17 Triangles

### Questions

1 A student asks you if there is such a thing as equivalent triangles. What is your reply?

2 A student asks, "Is it possible for one triangle to have five parts (angles, sides) congruent to five parts of another triangle and still not have the two triangles congruent?" How do you reply?

3 A student asks you if there are any triangles in which the perimeter equals the area. What is your reply?

4 "If a leg and an acute angle of one right triangle are congruent to a leg and an acute

angle of another right triangle, then the triangles are congruent." Do you agree with this student's statement?

5 A student in your class asks the following questions.

(a) "If $\triangle ABC \cong \triangle DEF$, is $\triangle ABC \cong \triangle EDF$?" How do you reply?

(b) "If $\triangle ABC \sim \triangle DEF$, is $\triangle ABC \sim \triangle EDF$?" How do you reply?

Why or why not (in each case)?

6 After you have introduced congruent triangles in your geometry class, a student asks, "If we can write $6 + 4 = 10$, then if $\triangle ABC$ is congruent to $\triangle DEF$, why can't we write $\triangle ABC = \triangle DEF$?" How would you answer?

7 A student asks, "Why is the triangle inequality so named?" How do you reply?

8 A student asks, "The two triangles below are congruent and you said that we must write it a certain way, i.e., $\triangle ACB \cong \triangle EGF$. If congruent triangles mean same size and shape, then why can't we write $\triangle ACB \cong \triangle GFE$ since triangle GFE is the same triangle as EGF, i.e., the same set of points?" How do you respond?

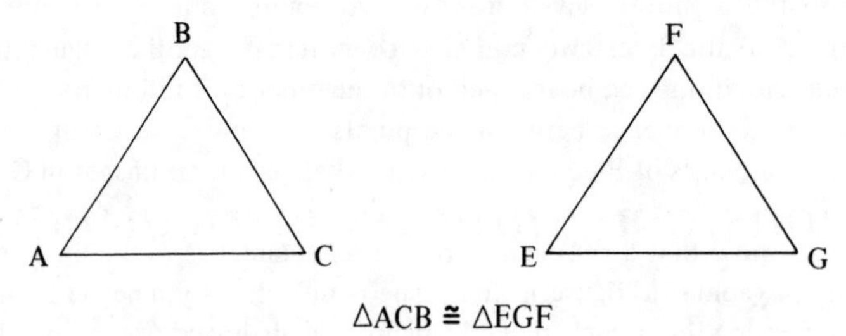


$\triangle ACB \cong \triangle EGF$

9 You have asked a student to name the shortest segment in the following figure. The student answers, "$\overline{AC}$ because it's opposite the smallest angle ($\angle CBA$ is the smallest angle in the figure)." Is the student correct?

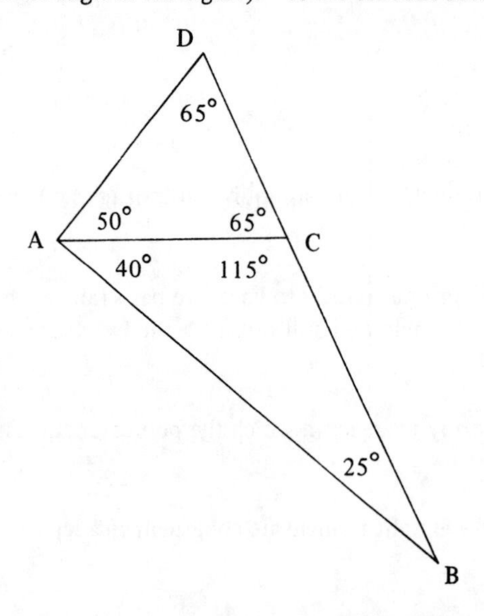

10 A student asks, "Does the segment containing the midpoints of two sides of a triangle have a name?" How would you answer?

11 A student asks you if the bisector of an angle of a triangle also bisects the opposite side. How do you respond?

## Answers

1 Equivalent triangles are frequently defined to be triangles that have the same area. Thus if $\Delta ABC$ has an area of 10 square units and $\Delta DEF$ has an area of 10 square units, then $\Delta ABC$ is equivalent to $\Delta DEF$.

***Problem*** Is the relation "is equivalent to" an equivalence relation on the set of triangles in a plane?

2 For an answer to this question, read:

R. Pawley, "5-Con Triangles," *The Mathematics Teacher*, May 1967, pp. 438-442.

3 For an answer to this question, read the following article by L. W. Smith.* Also consult the references given in Section 1.6(B), Answer 4.

# *Conditions governing numerical equality of perimeter, area, and volume*

LEANDER W. SMITH, *The University of Oklahoma, Norman, Oklahoma.*

*Development of theorems leading to pattern generalization*

UNDER WHAT CONDITIONS will the area and the perimeter of a triangle be numerically equal? Since the right triangle affords us some computational simplicity, let us consider the class of right triangles first.

Given right triangle $ABC$, with sides $a$ and $b$ and hypotenuse $c$, having area and perimeter numerically equal, we may write:

$$a^2+b^2=c^2, \quad (1)$$

$$a+b+c=\text{perimeter}, \quad (2)$$

$$\tfrac{1}{2}ab=\text{area}. \quad (3)$$

By equating (2) and (3), we obtain

$$a+b+c=\tfrac{1}{2}ab \quad (4)$$

and since

$$c=\sqrt{a^2+b^2},$$

we can rewrite (4) in two variables as

$$a+b+\sqrt{a^2+b^2}=\tfrac{1}{2}ab \text{ for some } a \text{ and } b. \quad (5)$$

*From *The Mathematics Teacher*, April 1965, pp. 303-307. Reprinted by permission.

Then,

$$\sqrt{a^2+b^2}=\tfrac{1}{2}ab-(a+b)$$
$$4(a^2+b^2)=(ab-2[a+b])^2$$
$$4a^2+4b^2=a^2b^2-4ab(a+b)+4a^2$$
$$+8ab+4b^2$$
$$0=a^2b^2-4ab(a+b)+8ab$$
$$0=ab-4a-4b+8$$
$$0=a(b-4)-4(b-2)$$

so

$$a=\frac{4(b-2)}{b-4},\quad b\neq 4. \qquad (6')$$

We also know that $a>0$, $b>0$, and that $a+b>c$ by the assumption of triangularity. By inspection we can see that $a>0$ for values of $b>4$, and that $a<0$ for $2<b<4$. It would appear that for $0<b<2$, $a>0$. However, knowing that $2c>0$ and from (5) that $2c=ab-2(a+b)$ we find

$$ab>2a+2b$$
$$ab-2a>2b$$

so

$$a>\frac{2b}{b-2},\quad b\neq 2. \qquad (5')$$

From the above conclusion, we see that $0<b<2$ would permit $a<0$, which is a contradiction. Combining the results of (5′) and (6′), we can now write

$$a=\frac{4(b-2)}{b-4} \text{ for all } b>4. \qquad (6)$$

By examining the graphic solution of (6) we shall find a clue to motivate a more general solution to the problem of having area and perimeter numerically equal. Consider the $X$, $Y$ plane and allow $x \in X$ and $y \in Y$ such that a function $G$ maps $X$ onto $Y$ by the following rule:

$$G\colon x \to \frac{4(x-2)}{x-4} \text{ for all } x>4.$$

By projecting each point $(x, y)$ of $G$ to the axes and joining the intercepts, we generate the set of right triangles from which we obtained (6). (See Fig. 1.)

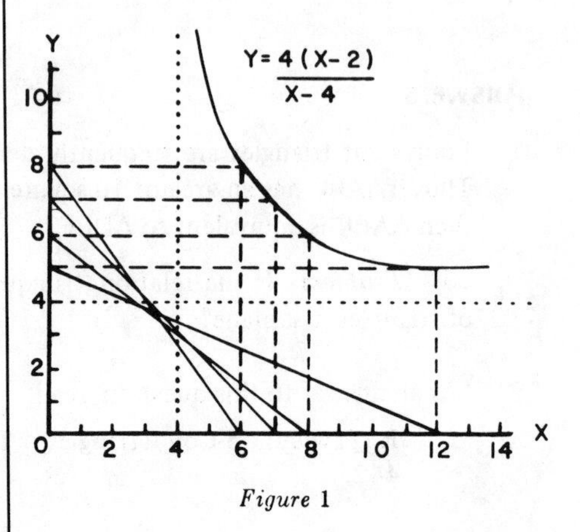


*Figure 1*

It should be noted that an isosceles right triangle is formed when $x=y=4+2\sqrt{2}$, the projection of the vertex of the curve. It can be seen that as $x$ increases without bound, $y$ approaches 4; moreover, the interval $(4, 4+2\sqrt{2}]$ is the set of all points whose coordinates represent the possible lengths of the short sides of right triangles, for which area and perimeter are numerically equal. Since 4 is not in the domain of $y$, there are exactly four integral solutions of $G$, namely, (5, 12), (6, 8), (8, 6), and (12, 5); thus, there are exactly two right triangles having sides of integral measure and having area and perimeter numerically equal. These are the 6-8-10 and 5-12-13 triangles.

Now, then, let us turn our attention to the set of hypotenuses in Figure 1. Looking at the drawing, it appears that the set of hypotenuses form a set of tangents to a circle of radius 2. This leads us to the solution for the general triangle.

Let us draw a scalene triangle $ABC$ with its incircle (center at O). Note that we may now join the incenter O to each vertex, thus partitioning triangle $ABC$ into three triangles $AOB$, $BOC$, and $COA$. (See Fig. 2.) Each of these triangles has as its altitude the radius of the incircle. Then, assuming triangle $ABC$ has its area

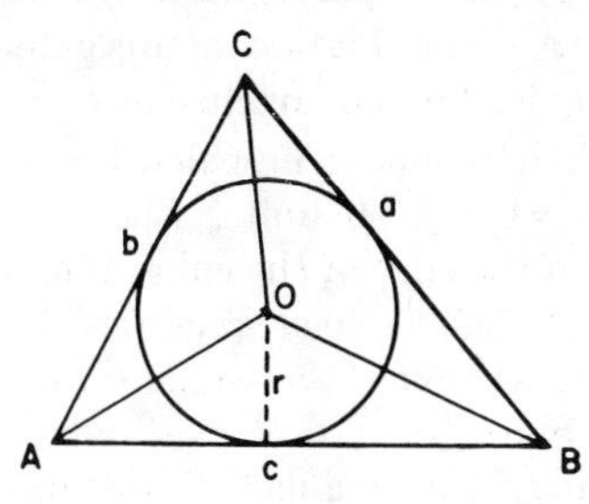


*Figure 2*

and perimeter numerically equal, we may write:

$$\tfrac{1}{2}ra+\tfrac{1}{2}rb+\tfrac{1}{2}rc=a+b+c \qquad (7)$$

or

$$\tfrac{1}{2}r(a+b+c)=a+b+c$$

hence,

$$r=2.$$

*Theorem* I′

If the area and the perimeter of a triangle are numerically equal, the radius of the incircle is two.

Since the proof of the converse of Theorem I′ is trivial, we may use the more powerful theorem below.

*Theorem* I

The area and the perimeter of a triangle are numerically equal *if and only if* the radius of the incircle is two.

Many mathematicians will argue that the result is trivial, since, by proper choice of the unit of measure, the inradius can be made to be two units. But to the high school student, the problem has interesting motivational properties and even more useful practice in applying the algebraic properties of the field. Students can also see the production of extraneous roots from (5) to (6) and are forced to think about the order properties on inequalities in the handling of (5′) and (6). What follows should afford even greater motivation and student level research.

If any polygon has an incircle of radius two, the area and the perimeter of the polygon are numerically equal. Is this immediately obvious? If not, draw an irregular polygon about a circle $O$. Then join $O$ to each vertex and note that the radius of the incircle is the altitude of each triangle formed. (Teachers of geometry will recognize that for regular polygons, the area is often given as one half the product of the perimeter and the apothem.)

Hence, we need not limit our discussion to triangles only. We can increase the number of sides of the polygon without bound, making sure that each side is tangent to the incircle. What happens? As the number of sides increases, the polygon approaches the incircle and its area approaches that of the circle. This seems like a good time to provide an intuitive notion of limit for, sure enough, as the number of sides increases without bound, a limit is being approached.

$$\pi r^2=2\pi r \text{ when } r=2.$$

What is the area of a square having an incircle of radius two?

Some polygons do not have incircles and yet have area and perimeter numerically equal. For example, rectangles with length ($l$) and width ($w$) where $l=\dfrac{2w}{w-2}$ and $w>2$ satisfy this condition. How many rectangles having sides of integral length are there? More generally, the set of all parallelograms for which the area and the perimeter are numerically equal may be given by

$$a=\frac{2b}{b\sin\alpha-2}$$

where $a$ and $b$ are the sides of the parallelogram and $\alpha$ is the included angle.

For those whose background includes an introduction to the calculus, a bit of generalization follows. Let a square have sides of measure $x$. Its area is then given by the formula $A=x^2$ and its perimeter by $p=4x$. Using the calculus produces some interesting results:

If $A=x^2$, then $\frac{dA}{dx}=2x$. Since $p=4x$, we may write $p=2\frac{dA}{dx}\cdot$ But when the area and the perimeter of a square are numerically equal, the length of a side, $x$, is four units and we could rewrite the expression for $p$ as

$$p=\frac{x}{2}\cdot\frac{dA}{dx}\cdot \tag{8a}$$

Examine another case, that of the equilateral triangle having area and perimeter numerically equal. This occurs when the length of a side is $4\sqrt{3}$ units. If the area is given by the formula $A=\frac{s^2\sqrt{3}}{4}$ when the perimeter is given by $p=3s$, $\frac{dA}{ds}=\frac{s\sqrt{3}}{2}\cdot$

So, if $p=3s$, we may write

$$p=4\sqrt{3}\ \frac{dA}{ds}$$

or, more compactly,

$$p=\frac{s}{2}\cdot\frac{dA}{ds}\cdot \tag{8b}$$

Do you see a pattern emerging from (8a) and (8b)? Try examining a circle in the same way. Sure enough, the result generalizes:

*Theorem* II

If the area ($A$) and the perimeter ($p$) of a polygon are numerically equal, then

$$p=\frac{x}{2}\ \frac{dA}{dx}$$

where $x$ is a linear measure on $A$ and $p$.

Let us turn our attention to a similar problem in three-space: Under what conditions are the volume ($V$) and the surface area ($S$) of a polyhedron numerically equal? An examination of discrete cases will help lead to our conclusion.

Begin with a cube in which $V=S$. Then since $s^3=6s^2$, $s=6$ units, where $s$ is the length of the edge of the cube. What is the radius of the inscribed sphere?

Next, examine the sphere for which $V=S$. Since $\frac{4}{3}\pi r^3=4\pi r^2$, $r=3$ units.

Finally, the regular tetrahedron for which $V=S$ has this relationship stated by $\frac{s^3\sqrt{3}}{12}=s^2\sqrt{3}$, whence $s=12$ units. The insphere in this case has a radius of three units.

Since there are polyhedra for which there is no insphere, we state a weak theorem, like Theorem I′, namely:

*Theorem* III

If a polyhedral $n$-gon has an insphere of radius three, the volume ($V$) and the surface area ($S$) are numerically equal.

A proof of Theorem III may be based on the fact that for any pyramid, the volume is one-third of the product of the base area ($B$) and altitude ($h$) of the pyramid. Hence, when the altitude is three units, the volume and surface area are numerically equal. (Bear in mind that the polyhedron we are describing may be thought of as a set of planes tangent to an insphere.)

Resorting to the calculus again, we discover an interesting relationship. In the case of the cube, with edge $e$, if $V=e^3$ then $\frac{dV}{de}=3e^2$; since we are concerned with the case when $V=S$, we find that $e=6$, so

$$S=\frac{e}{3}\ \frac{dV}{de}\cdot \tag{9a}$$

In the case of the tetrahedron, if $V=\frac{e^3\sqrt{3}}{12}$ then $\frac{dV}{de}=\frac{e^2\sqrt{3}}{4}$, and when $V=S$,

$$S=\frac{e}{3}\cdot\frac{dV}{de}. \quad (9b)$$

The sphere itself (as a limiting case) is also of interest. If $V=\frac{4}{3}\pi r^3$ and $S=4\pi r^2$ then $\frac{dV}{dr}=4\pi r^2$, and when $V=S$, $r=3$ and

$$S=\frac{r}{3}\cdot\frac{dV}{dr}. \quad (9c)$$

Again we are about to generalize from (9a), (9b), and (9c).

*Theorem IV*

If the volume ($V$) and the surface area ($S$) are numerically equal, then

$$S=\frac{x}{3}\,\frac{dV}{dx}$$

where $x$ is a linear measure on $V$ and $S$.

"Aha," you say, "in two-space $p=\frac{x}{2}\,\frac{dA}{dx}$ and in three-space $S=\frac{x}{3}\,\frac{dV}{dx}$!"

One of the beauties of mathematics is the persistence of pattern generalization. What happens if we extend this problem to $n$-space and talk about the volume ($V$) and surface area ($S$)? Let's see.

If $V=c_1x^n$ and $S=c_2x^{n-1}$, then we see that if $V=S$, $x=\frac{c_2}{c_1}$. Using the calculus, $\frac{dV}{dx}=nc_1x^{n-1}$ and we can express $S$ in terms of $\frac{dV}{dx}$ as follows:

$$S=\frac{c_2(nc_1x^{n-1})}{c_1n}.$$

The parentheses enclose what we recognize to be $\frac{dV}{dx}$ and $\frac{c_2}{c_1}=x$ when $V=S$; so our equation becomes $S=\frac{x}{n}\cdot\frac{dV}{dx}$. Hence, we have a general theorem which summarizes the results of Theorem II and Theorem IV, which follows.

*Theorem V*

In $n$-space, if the volume and the surface area are numerically equal, i.e., $V=c_1x^n$ and $S=c_2x^{n-1}$, then

$$S=\frac{x}{n}\,\frac{dV}{dx}$$

where $x$ is a linear measure on $V$ and $S$.

Further extensions remain to be explored. Meanwhile, the spontaneous generation of theorems resulting from conjectures in this paper provide a model of student-teacher directed classroom inquiry.

REFERENCES


COURT, NATHAN A. *College Geometry.* New York: Barnes & Noble, 1952 (2nd ed.).

COXETER, H. S. M. *Regular Polytopes.* New York: The Macmillan Company, 1963 (2nd ed.).

SMITH, LEANDER W. "A dialogue on Two Triangles," THE MATHEMATICS TEACHER, LVII (April, 1964), 233–34.


4 No. Consider Figure 3.17.1. $\angle A \cong \angle F$ and $\overline{BC} \cong \overline{EF}$, but $\triangle ABC$ is not congruent to $\triangle DEF$. A correct statement would be:

> If a leg and an acute angle of one right triangle are congruent to the corresponding parts of another right triangle, then the triangles are congruent.

*FIGURE 3.17.1*

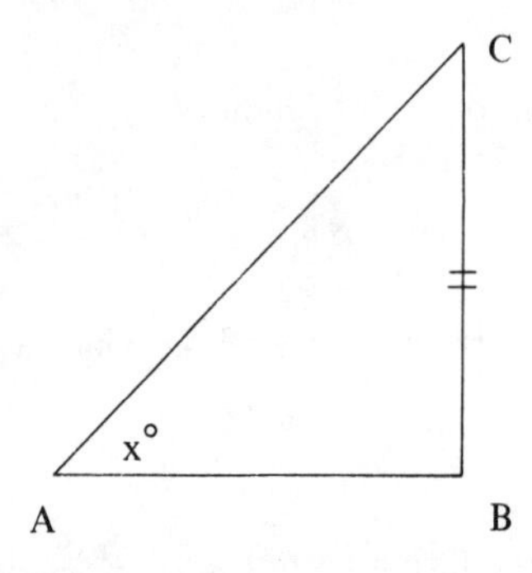


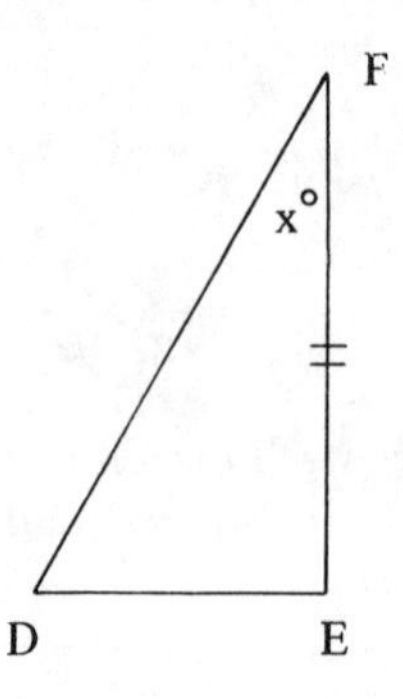


5 Part (a) is answered here, but (b) is left for the reader.

(1) Not necessarily. Since $\triangle ABC \cong \triangle DEF$, then by the convention adopted by many texts we know the following:

$\overline{AB} \cong \overline{DE}$ $\angle A \cong \angle D$
$\overline{BC} \cong \overline{EF}$ $\angle B \cong \angle E$
$\overline{AC} \cong \overline{DF}$ $\angle C \cong \angle F$

(2) If $\triangle ABC$ is congruent to $\triangle EDF$, then similarly it would follow that:

$\overline{AB} \cong \overline{ED}$ $\angle A \cong \angle E$
$\overline{BC} \cong \overline{DF}$ $\angle B \cong \angle D$
$\overline{AC} \cong \overline{EF}$ $\angle C \cong \angle F$

However, these latter congruences do not necessarily follow from the former congruences.

***Problem 1*** Under what conditions will the student's statement be true in (1).

***Problem 2*** Under what conditions will the student's statement be true in (2).

6 This problem presents a difficulty as to what the student is really asking. That is, what does $6 + 4 = 10$ have to do with the rest of the question? However, this question is worded exactly as it occurred in a high school classroom. It seems as if the student is asking that since one can have two different names for the number ten under equality, then why can't we have two different names for triangles under equality.

The notation $\triangle ABC = \triangle DEF$ is usually reserved to mean that these two triangles are the same set of points. The notation $\triangle ABC \cong \triangle DEF$ means, intuitively, that two triangles have the same size and shape. Thus two triangles that are congruent ($\cong$) do not necessarily have to have the same set of points.

7 If three points A, B and C are situated so that $AB + BC = AC$, then the point B is between A and C, and all three points lie on one line. If $AB + BC > AC$, and B is not on the same line as A and C, then the points A, B and C determine a triangle. A familiar theorem from geometry states that the sum of the lengths of any two sides of a triangle exceeds the length of the third side.

8 The notation that is usually used for congruent triangles, $\Delta ACB \cong \Delta EGF$, was selected so that one could tell easily which parts in one triangle correspond to which parts in another triangle. Suppose we are given the situation shown in Figure 3.17.2. Then we know that $\angle A \cong \angle E$, $\angle B \cong \angle F$ and $\angle C \cong \angle G$. Also $\overline{AB} \cong \overline{EF}$, $\overline{BC} \cong \overline{FG}$ and $\overline{AC} \cong \overline{EG}$. All of this information is given to us by using the suggested notation, i.e. $\Delta ACB \cong \Delta EGF$.

*FIGURE 3.17.2*

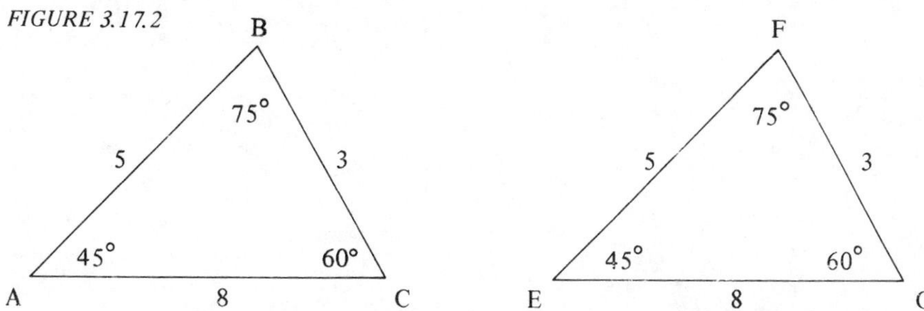


If we were to also write that $\Delta ACB \cong \Delta GFE$, then with the notational convention cited above, this would imply that, for example, $\overline{AC} \cong \overline{GF}$ and $\angle A \cong \angle G$, etc., which is certainly not true.

A congruence between triangles is often defined as follows.

If a correspondence between the vertices of two triangles is such that each side and each angle of one triangle is congruent to the corresponding part of the other triangle, then the correspondence is called a congruence between the triangles.

Thus in the above, the correspondence $\Delta ACB \leftrightarrow \Delta GFE$ is not a congruence.

Thus even though $\Delta EGF$ and $\Delta GFE$ are the same triangle, i.e., the same set of points, in order to show that two triangles are congruent, one must be careful to show which parts of one triangle correspond to which parts of the other triangle.

9 No, the student is not correct. The shortest segment in the figure is $\overline{DC}$.

***Problem 1*** Write out a detailed explanation showing that the above statement is true.

***Problem 2*** State and prove the main theorem that you used to justify your assertions in Problem 1.

10 In some texts, this segment is given the name *midline* while in other texts no name is given.

11 The following statement is usually proved as a theorem or given as an exercise in many geometry texts.

If a ray bisects an angle of a triangle, it divides the opposite side into segments whose lengths are proportional to the lengths of the other two sides.

***Problem 1*** Accepting the above theorem, what example(s) would you give to show the student that his statement is not necessarily true? Under what circumstances would the student's statement be true?

***Problem 2*** Prove the theorem given in the answer above. (*Hint* Use parallel lines and a transversal.)

# Index

This index contains references to both Part 1 and Part 2 of the set.